THE ARTS OF 17TH-CENTURY SCIENCE: REPRESENTATIONS OF THE NATURAL WORLD IN EUROPEAN AND NORTH AMERICAN CULTURE

The Arts of 17th-Century Science

Representations of the natural world in European and North American culture

Edited by

Claire Jowitt and Diane Watt

Routledge
Taylor & Francis Group
LONDON AND NEW YORK

First published 2002 by Ashgate Publishing

2 Park Square, Milton Park, Abingdon, Oxon OX14 4RN
711 Third Avenue, New York, NY 10017, USA

Routledge is an imprint of the Taylor & Francis Group, an informa business

First issued in paperback 2016

British Library Cataloguing in Publication Data

The arts of 17th-century science : representations
of the natural world in European and North American culture
1. English literature – Early modern, 1500–1700 – History
and criticism 2. American literature – Colonial period, ca.
1600–1775 – History and criticism 3. Science in literature
4. Nature in literature
I. Jowitt, Claire II. Watt, Diane
820.9'356'09032

Library of Congress Control Number: 2002108550

ISBN 978-0-7546-0417-4

Transferred to Digital Printing in 2011

ISBN 978-0-7546-0417-4 (hbk)
ISBN 978-1-138-25643-9 (pbk)

Contents

List of Figures

Acknowledgements

This book evolved from the conference on Science and Literature in the Seventeenth Century, which was held in Aberystwyth in the summer of 1997. Our first acknowledgement, then, is to all the speakers who made that weekend such an enjoyable and stimulating event. In organising the conference and in editing this volume, we accumulated a significant number of other debts. To our colleagues (past and present) and our friends, who either attended the conference, chaired sessions, or offered their time and support in a myriad of ways, all we can do is say thank you and emphasise how much we appreciate everything you have done. In particular, though, we would like to thank Andrew Hadfield, Jonathan Sawday, Sarah Prescott and Shaun Regan, who, in their different ways, made substantial contributions. For institutional support and financial assistance we gratefully acknowledge Brean Hammond, Lyn Pykett and Steve Smith. We would like to express our appreciation of Erika Gaffney, of Ashgate Publishing Company, whose enthusiasm for and faith in the book has been so encouraging. Andrew Hadfield read and commented upon the introduction to the volume, as did Martin Willis, and we are indebted to them for their suggestions; all errors that remain are, of course, our own. We are also indebted to the readers appointed by the publishers for such informed and constructive advice, and to the contributors themselves, for their patience and hard work. Finally, a special debt of thanks is owed to Heike Bauer, who played an invaluable role in the production of the final manuscript.

Parts of this book first appeared in earlier publications, although they have since been revised. Bronwen Price's essay originally appeared in the journal *Literature and History*, **7** (1998); it is used here by permission of the editors and Manchester University Press. Mary Baine Campbell's essay is reprinted from 'On the Infinite Universe and Innumerable Worlds', in *Wonder and Science: Imagining Worlds in Early Modern Europe* (1999, Cornell University) and is used here by permission of the publisher, Cornell University Press. Despite every effort to trace and contact copyright holders prior to publication, this may not have been possible in every case. We apologise for any apparent infringement of copyright and, if notified, we will be pleased to rectify any errors or omissions at the earliest opportunity.

This book is dedicated to Carola Scott-Luckens (1946–2002). As an undergraduate, a research student and a post-doctoral fellow at the University of Southampton, Carola contributed to the debates about early-modern literature and culture held there for 10 years. We are privileged to include her excellent essay on millenarian utopias and Baconianism in this volume. She will be missed as a fellow researcher but also as a valued friend by many of the people included in this book.

Notes on Contributors

Anthony R. Archdeacon is an independent scholar, with research interests in medieval and renaissance literature and ideas; his doctoral thesis investigated the idea of nothingness in the Renaissance. He taught at Southampton University and Southampton New College, and is currently Principal Officer for English at the Qualifications and Curriculum Authority. He also works on ways of integrating information and technology into the teaching of English.

Andrew Bradstock is Secretary for Church and Society at the United Reformed Church in London, and has previously held lecturing and research posts in the departments of Theology at King Alfred's College, Winchester; University of Southampton New College; La Sainte Union College of Higher Education, Southampton and the University of Otago, Dunedin. His publications on Winstanley include *Faith in the Revolution: The political theologies of Müntzer and Winstanley* (1997), and, as editor, *Winstanley and the Diggers, 1649–1999* (2000). He has also written widely on political and liberation theology, and co-edited two books on Victorian spirituality and sexuality.

Mary Baine Campbell is Professor of English and American Literature at Brandeis University, where she teaches medieval and early modern literature, as well as the writing of poetry. She is the author of *The Witness and the Other World: Exotic European Travel Writing, 400–1600* (1988) and the recent *Wonder and Science: Imagining Worlds in Early Modern Europe* (1999), as well as two books of poems, *The World, the Flesh, and Angels* (1989) and *Trouble* (2002). Her current work in early modern studies concerns itself with dreams both private and publicly broadcast, especially but not exclusively the dreams of natural philosophers. She is also writing the libretto, from Marie de France's twelfth-century lai, the *Bisclavret*, for an opera by Martin Brody.

Jess Edwards lectures in Early Modern English Literature at the University of North London. He has published essays, and is currently completing a book, on mathematics, property and colonisation in seventeenth-century England and America.

Ruth Gilbert is Lecturer in English at the University of Southampton. She is co-editor (with Erica Fudge and Sue Wiseman) of *At the Borders of the*

Human: Beasts, Bodies and Natural Philosophy in the Early Modern Period (1999). Her book *Early Modern Hermaphrodites: Sex and Other Stories* was published by Palgrave in 2002.

Andrew Hadfield is Professor of English at the University of Wales, Aberystwyth. He is the author of a number of studies of Renaissance literature and culture, including *Spenser's Irish Experience: Wilde Fruit and Salvage Soyl* (1997), *Literature, Travel, and Colonial Writing in the English Renaissance, 1545–1625* (1998) and *The English Renaissance, 1500–1625* (2000). He has recently edited *The Cambridge Companion to Spenser* (2001) and an anthology, *Amazons, Savages and Machievals: Travel and Colonial Writing in English, 1550–1630* (2001). He is general editor of the Arden Critical Companions and his *Shakespeare and Politics* will be published in 2003. He is a regular reviewer for the *Times Literary Supplement.*

Elaine Hobby is Professor of Seventeenth-Century Studies at Loughborough University. She has been researching women's writing from the period 1640–1700 since 1978, and finds it more fascinating as each year passes. Her publications include: *Virtue of Necessity: English Women's Writing 1649–1688* (1988); co-editorship, with Elspeth Graham, Hilary Hinds and Helen Wilcox, of *Her Own Life: Autobiographical Writings by Seventeenth-Century Englishwomen* (1989); her edition of Jane Sharp, *The Midwives Book* (1999). Inspired by Jane Sharp's wit, she is currently working on a history of the early-modern midwifery manual, and gratefully acknowledges the support given to this project by a grant from the Wellcome Trust.

Claire Jowitt teaches English Renaissance literature at the University of Wales, Aberystwyth. Her book, *Voyage Drama and Gender Politics 1589–1642: Real and Imagined Worlds*, will be published by Manchester University Press (forthcoming 2002). She has published a range of essays on women's writing, the colonisation of America, and Jewish–Christian relations in the seventeenth century. She has recently edited a special issue of the journal *Women's Writing* on *Dissenting Women 1350–1800.* She is currently working on a book on Jews, Turks and pirates on the early-modern Stage.

Bettina Mathes teaches cultural history at Humboldt University. Her research favours transdisciplinary approaches and focuses on the history of gender, sexuality and the media. She is currently working on a book, tentatively titled *Virtual Regenerations*. Her publications include *Verhandlungen mit Faust. Geschlechterverhältnisse in der Kultur der frühen Neuzeit.* Königstein: Helmer (2001) and 'Doctor Faustus Impotent? Fantasizing the Male Body in the Historia von D. Johann Faustus', in *Women In German Yearbook 15*

(2000), and 'As Long As A Swan's Neck? The Significance of the "Enlarged" Clitoris for Early Modern Anatomy', in Elizabeth D. Harvey, ed. *Sensible Flesh: On Touch in Early Modern Culture* (University of Pennsylvania Press, 2002).

Peter Mitchell has recently completed a PhD at the University of Wales, Lampeter, entitled 'The Anatomical Speaking Picture of *The Purple Island*: An Index to Anatomy in Early Seventeenth-Century Christian Literature, Natural Philosophy and Theology'. He also has research interests in theories of figurality, narrative, the human body and subjectivity.

Bronwen Price is Senior Lecturer in English at Portsmouth University. She specialises in seventeenth-century literature and has published a range of essays and articles on early-modern women's writing and seventeenth-century poetry. She has recently finished editing a volume of new essays on Francis Bacon's *The New Atlantis* for Manchester University Press (forthcoming, 2002).

Jonathan Sawday is Professor of English Studies, and Head of the Department of English at the University of Strathclyde in Glasgow. He is the author of *The Body Emblazoned: Dissection and the Human Body in Renaissance Culture* (1995), and co-editor (with Thomas Healy) of *Literature and the English Civil War* (Cambridge, 1990) and (with Neil Rhodes) of *The Renaissance Computer: Knowledge Technology in the First Age of Print* (Routledge, 2000).

Carola Scott-Luckens, following her doctoral dissertation on '*Women's Millennialist Prophecy 1630–1670*', continued to study early-modern millenarian writings and the extent to which expectations of a messianic second coming were interwoven with English and continental politics and philosophy during the mid-seventeenth century. Among her recent publications, 'The Broken Tabernacle', on the religio-political significance of the 1647 soul-narrative of Baptist visionary Sarah Wight by her pastor, is available online in Volume I of the *Chichester College of Higher Education Journal Signatures*. 'An *Instauratio Magna* of Universal Fellowship? Proposals for a Judaic University in Revolutionary London', examining the influence of Baconianism in utopian texts produced within the Hartlib circle in the 1640s and 1650s, features in the journal *Jewish Culture and History*, vol. 3, no. 1, 2000, pp. 75–95. She also contributed the entry on the Fifth Monarchist prophetess Anna Trapnel, in the Women Writers Project database published by Brown University. Earlier this year, after a long illness, Carola lost her battle against leukemia.

David E. Shuttleton teaches literature, theory and film at the University of Wales, Aberystwyth, and has research interests in medicine and early-modern literature, and gay fiction. He has published widely on late-seventeenth- and eighteenth-century medico-literary culture with a special interest in the physician George Cheyne (1672–1743). He contributed to Roy Porter (ed.), *Medicine in the Enlightenment* (1996), and Isobel Armstrong and Virginia Blain (eds), *Women's Poetry in the Enlightenment* (1999). He co-edited (with Richard Phillips and Diane Watt) *De-centering Sexualities: Politics and Representations Beyond the Metropolis* (2000). He is a contributing editor to the forthcoming *Cambridge Edition of the Works and Correspondence of Samuel Richardson*.

Richard Sugg is Lecturer in Early Modern English literature at Cardiff University. He is currently completing an article on the use of anatomical rhetoric in late Tudor and early Stuart England. From September 2002 he will be conducting an internally-funded research project on the Lenten 'anatomical season' of early seventeenth-century London. Other research interests include Renaissance body–soul physiology; the use of anatomical imagery during the English Revolution; anatomical and corporeal language in Protestant, and especially Puritan, literature; and the relationship between new world colonialism, evangelism, and apocalypticism.

Diane Watt is Senior Lecturer in English at the University of Wales, Aberystwyth. Her main research interests are in the fields of medieval and renaissance literature and history, and gender and sexuality studies. She is the author of *Secretaries of God: Women Prophets in Late Medieval and Early Modern England* (Brewer, 1997), editor of *Medieval Women in Their Communities* (University of Wales Press and University of Toronto Press, 1997) and co-editor (with Richard Phillips and David Shuttleton) of *De-centring Sexualities: Politics and Representations Beyond the Metropolis* (Routledge, 2000). She is currently working on a book on language, sex and politics in John Gower's *Confessio Amantis*.

Introduction

Claire Jowitt and Diane Watt

> So at length, when universal learning has once completed its cycle, the spirit of man, no longer confined within this dark prison-house, will reach out far and wide, till it fills the whole world and the space far beyond with the expansion of its divined greatness. Then at last most of the chances and changes of the world will be so quickly perceived that to him who holds this stronghold of wisdom hardly anything can happen in his life which is unforeseen or fortuitous. He will indeed seem to be one whose rule and dominion the stars obey, to whose command earth and sea hearken, and whom winds and tempests serve; to whom, lastly, Mother Nature herself has surrendered, as if indeed some god had abdicated the throne of the world and entrusted its rights, laws, and administration to him as governor.[1]

John Milton's optimistic assertion of the powers of 'universal learning' in *Prolusions* (1628–32; published 1674) is emblematic of the complexities of seventeenth-century views of the capacities and promise of natural knowledge. Earlier in the century, Francis Bacon's aphorism '*Plus ultra*' rather than '*ne plus ultra*' had encapsulated the ambition of the age in its representation of the scientist as hero,[2] and the Baconian influence on Milton is here evident. Furthermore, there is a radical political agenda apparent in this quotation. For Milton, the male natural philosopher becomes, as it were, the elected 'governor', even Protector, of the world.[3] This fantasy of bloodless revolution is made possible by the resignation of the 'god' or previous monarch from his 'throne'. Just as Jupiter replaced Saturn and ushered in a Golden Age, so new science will introduce a time of wisdom, justice and prosperity. In contrast, though, to the power struggle between Jupiter and Saturn for control of the classical world, and to the bitter rivalry between the Stuart monarchs and their Parliaments in the seventeenth century, here the monarch willingly 'abdicate[s]' the throne to welcome in a new, more popular, regime. Indeed the vicissitudes and uncertainties of early-modern existence will be rendered obsolete since new science will be able to predict, and hence control, the future. We have here, then, a conflation of mystical, prophetic rhetoric with empirical language and ideals. Furthermore, there is a powerful sense of anticipation within Milton's vision. He contemplates a supine, feminised, natural world colluding in her own submission to this virile new patriarchal authority. Similar to colonial propaganda of this period in which Britain attempted to invent itself as an Imperial power and to fashion, often imaginatively, an Empire, Milton imagines new science capable of 'command[ing] earth and sea'.[4] Implicit in Milton's representation of natural philosophy is an ethical concern with working for the greater good of society and of humankind as a whole. Despite the collaborative tenor of the early Royal Society (founded 1662), as expressed in Thomas Sprat's 1667 *History of the Royal Society*, Milton's new scientist remains resolutely solitary.[5] Solus, he stands contemplating the world before him. There is a palpable sense too that, as Bacon and Sprat had imagined in *The Advancement of Learning* and *History of the Royal Society* respectively, the new philosopher can restore man's lost dominion over nature.

The level of excited and breathless anticipation apparent in Milton's *Prolusion* is not maintained in his other works; nor is his confidence in the heroic endeavour of the new scientific project. John Rogers has persuasively demonstrated Milton's debt to vitalist philosophy (which, according to Roger's definition, 'holds in its tamest manifestation the inseparability of body and soul and, in its boldest, the infusion of all material substance with the power of reason and self-motion'[6]). However, Kester Svenden and Stanley Fish, in particular, have argued that in *Paradise Lost* (published in 1667, but thought to have been composed in 1658–60), Milton reveals his disquiet concerning experimental philosophy more generally.[7] According to such readings, in Milton's version of the Fall, Satan, then Eve and, ultimately, Adam are all influenced by empirical thought. Satan, the new scientist, persuades Eve of the efficacy of the experiment to eat the fruit from the Tree of Knowledge, and she in turn passes on her newly acquired knowledge to her partner Adam. It is this which results in the expulsion from Eden. Central to this interpretation are Eve's words celebrating experience which she utters after having consumed the forbidden fruit:

> Experience, next to thee I owe,
> Best guide; not following thee, I had remained
> In ignorance, thou open'st wisdom's way,
> And giv'st access, though secret she retire.[8]

As Karen Edwards has recently observed, '[Eve] plucks fruit from a tree and consumes it, at the behest of a serpent, in a paradisal garden. In the context of this depiction of the natural world, the term *experience* unmistakably gestures towards the new, or experimental, philosophy.'[9] There is certainly evidence to support the contention that Milton was sceptical about the benefits of these developments. For example, as Ruth Gilbert reveals in Chapter 10 of this volume, Satan's appropriation of the female role in reproduction as he gives birth to Sin in Book 2 of *Paradise Lost* contrasts with Adam's painless parturition described in Book 8, and hence disrupts a univocally positive view of the potentialities of new science. According to Gilbert, 'In representing these original moments of male reproduction Milton suggests the limits as well as the possibilities of masculine creation, procreation and (re)creation. Early-modern representations of male autogenesis could only ever flirt with the fantasy of erasing the female presence in reproduction.'[10] The problem that Milton struggles with here is that experiments, and experimental philosophy, take place in a fallen world. Indeed, his scepticism is apparent through the fact that, even before the Fall, Nature was out of control; the prelapsarian garden was becoming overgrown despite the best efforts of Adam and Eve ('for much their work outgrew / The hands dispatch of two gard'ning so wide').[11] Just as Milton fantasises about marginalising women within reproduction but keeps retreating from the idea, there is also a sense in which he questions whether science has any kind of power to restore Nature to a prelapsarian condition since that perfection never existed. Bacon's vision in *New Atlantis* that science will lead to 'the effecting of all things possible' – including the Restoration of Nature – might be seen to be treated by Milton rather more equivocally.[12]

Yet, Milton's attitude to the new philosophy as personified by Satan and Eve is rather more complex than either Svendsen or Fish allows.[13] For instance, in Fish's view Eve was culpable when she ate the fruit of the Tree of Knowledge because she betrayed her own innate powers of understanding and reasoning. She should have realised the monstrousness of Satan's suggestion that the fruit was responsible for giving a mere serpent the ability to speak. Hence, given the evidence before her of the silent animals and her failure to see Satan himself perform the act of consumption (in other words, to view the experiment), she should have doubted the accuracy of his claim 'whoso eats therof, forthwith attains / Wisdom' (*Paradise Lost*, IX.591–2). Karen Edwards interprets Eve's failure as that of a 'budding natural philosopher',[14] who does not pursue her own doubts about the serpent's assertions or insist on testing their validity. Yet this analysis is not taken to its logical conclusion. In her reading of this part of Milton's text, Edwards does not fully explore the significance of Eve's gender. Eve's isolation *as woman* in a masculine and patriarchal world needs to be considered. In *Paradise Lost* Eve works independently of Adam in the garden, and when the angel Raphael spoke to Adam of the likely attempt of Satan against the garden and its inhabitants Eve only 'overheard' the conversation since, she tells her husband, 'in a shady nook I stood behind'.[15] Bereft of a community of intellectuals, uninitiated in scientific discourse, her ability to experiment empirically and to pursue intellectual enquiry is flawed from the outset. In contrast to the first Fellows of the Royal Society, she is condemned to experiment alone without proper guidance and advice. It is no surprise that she is impressed by Satan's magic and distracted from her initial empirical questions. A parallel may be drawn here with the visit of Margaret Cavendish (one of the first female scientists) to the Royal Society in 1667, where instead of being shown experiments designed to demonstrate something new, she was merely entertained with the spectacular but pedestrian sight of watching a piece of meat being dissolved in acid – in effect, a sleight of hand or magic trick.[16] The particular difficulties faced by a female scientist, and Cavendish's responses to those gender-related problems in *The Blazing World* (1666), are explored in Bronwen Price's essay in this volume (see Chapter 8).

It is crucial not to view the workings of the Royal Society as upholding a distinction between the learned and the superstitious, the rational and the mystical at this time. According to Edwards, Satan's 'wisdom' is a perversion of Royal Society practice.[17] Satan lies to Eve and consequently it is not the experimental method but the scheming and deceptive individual practitioner that is culpable here. Satan's findings are fraudulent. Yet, Edwards also proposes that the values which Satan represents and the arguments he offers are not a straightforward reflection on those promulgated by practitioners of the early Royal Society. Rather, they stand for an intermingling of new and old philosophy. According to this view, Milton's depiction of Satan is indebted to the notion of the magus, and his transformation into talking serpent is the outcome of his dangerous pastime of dabbling in occult practices. However, Edwards here ascribes to the teleological narrative which views the scientific revolution as a fundamental disjunction from previous conceptions of the natural world. In other words, by arguing that Satan's occultism is oppositional to the practices of experimental philosophy, she elides the prominence of cultural conflicts inherent in the understanding of natural knowledge at this time. It is reductive and misleading to claim that it is Satan's occultism that undermines his status as an experimental philosopher. As Roy Porter outlines 'all manner of rival knowledge-claims … and, behind these … rival ideologies, promoted by disparate interests, such as religious, national and factional' co-existed in the early modern period, which led to 'multiple models of epistemological legitimacy'.[18] To put it another way, even important 'scientific discoveries' such as Isaac Newton's theories of attraction were informed by, rather than oppositional to, Newton's alchemical pursuits.[19] The 'mystical' and the 'scientific' are invariably both present in the work of a single author. As Giovanni Battista della Porta of Naples explained in *Natural Magick* (translated from the Latin in 1658), the study of magic was not at odds with empirical pursuits, rather such observational studies sought to 'survey the whole course of nature'.[20]

Newton's theories and discoveries were, of course, only a small part of the innovative scientific experiments being conducted in the seventeenth century. The central importance of understanding the work of the seventeenth-century natural philosophers in their broad cultural context is signalled in Charles Webster's seminal study of Puritan scientific endeavour, *The Great Instauration*:

> We have been misled by the subtle philosophical reasonings of Hobbes, the rigorous experimentalism of Harvey or Boyle, and the abstract mathematical analysis of Newton, into believing that the natural philosophers before 1700 thought in essentially modern terms, differing merely in lacking our considerable resources of experience and information. Full exploration of the evidence indicates that any understanding of the seventeenth-century worldview necessitates attention to the character of religious motivations, of adherence to philosophical tenets which have subsequently been discarded, and of involvement in contemporary political affairs. In other words, while a number of principles were adopted in the seventeenth century which are congruent with modern scientific knowledge, the general conceptual framework in which these principles were maintained was entirely different.[21]

Twenty-first century ideals of science representing disinterested and objective fields of investigation may have had their origins in the seventeenth century, but the late Renaissance was a period in which understandings of the natural world were vigorously contested. A key aim of this volume is to capture and trace the dynamic belief systems of early-modern society and to examine the ways in which discourses of science are implicated in and collapse into other cultural forms, and vice versa. Until recently, histories of science in the period have tended to argue that the plain and rational style and methodology of the Baconian empiricist Royal Society eclipsed earlier metaphysical, spiritual, imaginative or literary models. By the end of the seventeenth century, so the argument goes, certainties of religious belief and inherited tradition were increasingly at odds with the new 'cultures of discovery' which sought continuously to push forward the boundaries of knowledge and human understanding.[22] However, empiricism did not simply or uniformly replace Neoplatonist or spiritual beliefs, but entered into competition with them. Steve Shapin argues that there was 'a diverse array of cultural practices aimed at understanding, explaining, and controlling the natural world, each with different characteristics and each experiencing different modes of change'.[23] It is these complex processes of competition and negotiation concerning ways of seeing the natural world that are charted by the essays in this book. The collection explores the many overlaps between 'literary' and 'scientific' discourses as writers in this period attempted to understand, both imaginatively and empirically, the workings of the natural world, and shows that a discrete separation between such spheres and discourses is untenable. As Thomas Browne put it: 'There is in these workes of nature, which seeme to puzle reason, something Divine, and hath more in it than the eye of a common spectator doth discover.'[24] Browne's combination of argument from experience, religious and alchemical belief, and poetic language, here does not represent a discernible stance in relation to empiricist/spiritualist or rational/imaginative divides, but rather exemplifies the conflicts and compromises between them. The essays in this volume focus on the contests between these different ways of seeing and understanding the natural world in a wide range of European and North American texts from the period 1600–1700: in poetry and art, in political and religious writings, in descriptions of real and imagined colonial landscapes, as well as in more obviously 'scientific' documents. As Webster stresses, 'any understanding of the seventeenth-century worldview necessitates attention to the character of religious motivations, or adherence to philosophical tenets which have subsequently been discarded, and of involvement in contemporary affairs'.[25] The interdisciplinary scope of this volume is, then, crucial to the fulfilment of its aim of recovering the cultural as well as conceptual framework of the new science. Our endeavour is, of course, indebted to such ground-breaking studies as Steve Shapin and Simon Schaffer's *Leviathan and the Air Pump*, Denise Albanese's *New Science, New World*, and Lorraine Daston's *Wonders and the Order of Nature*.[26]

After Jonathan Sawday's contribution concerning the continued interaction between modern and seventeenth-century conceptions of science and natural knowledge, the volume is subdivided into four sections which reflect the major cultural concerns of the period. Part I, 'Philosophy, Thought and Natural Knowledge' engages with some of the intellectual ideas in circulation in the period. Anthony Archdeacon's essay (Chapter 2) interrogates the relationship between material and immaterial realities as examined in philosophical discourse, and its influence on the literature of the time, and especially the poetry of John Donne, Sir John Davies, William Shakespeare, Thomas Traherne and Cavendish. Focusing on the discovery of the vacuum, which was demonstrated by experiment to have actuality rather than to be only imaginable, and on the concept of 'non-beings', it examines the impact of Cartesian ideas on poetic notions of existence and fiction. In Chapter 3, Jess Edwards considers the relationship between the ideal and the material in late sixteenth- and early seventeenth-century discourses on mathematics (especially prefaces, lectures, speeches, anecdotes and metaphors). These discourses, which attempted to define the meaning of the subject, may, Edwards argues, give us greater insight into early-modern culture. Edwards notes that the continued influence of Neoplatonism was not (necessarily) antithetical to early-modern mathematical thinking. He contrasts the idealism of the English 'magus' John Dee with the antimystical empiricism of his friend and correspondent, the Huguenot martyr Petrus Ramus, and draws comparisons with Thomas Hood's English translation of Ramus. Edwards observes that apparently distinct schools of thought could share philosophical debts as well as common ambitions and goals. The last chapter in this section moves on a century, to Scotland on the eve of the Enlightenment. Here, David Shuttleton explores the charges of irreligion surrounding the controversial career of the Edinburgh physician, anatomist, poet and virtuoso, Dr Archibald Pitcairne. Pitcairne was a lifelong Jacobite, who participated in the divisions fostered by William of Orange's establishment of Presbyterianism. In the charged, sectarian climate of 1690s Edinburgh, all parties wielded the potentially capital charge of irreligion and atheism. This essay draws on Pitcairne's own writings and places them alongside neglected topical poetry, legal depositions and diaries. Shuttleton addresses the perceived tensions between Pitcairne's political conservatism and his potentially heretical, proto-Enlightenment intellectual project.

Part II addresses the interaction of theological, religio-political and scientific developments. Peter Mitchell's essay (Chapter 5) explores the connection between anatomical representation and Catholic belief, particularly eschatology and Eucharistic theology. It focuses on the relationship between the late sixteenth-century anatomical illustrations of Charles Estienne's 'De Dissectione' and the iconography of paintings of the sagittation of St Sebastian and his cure by St Irene in Counter-Reformation Italy, France and Spain. Mitchell proposes that the anatomical illustrations constitute an attempt to manipulate an already existing set of correspondences, in which the tools of the martyr's punishment can be understood as a means of medical recovery or bodily resurrection. Consequently, dissected criminal cadavers were iconographically transformed into the incorporeal bodies of redemptive sufferers, thereby deflecting from anatomists the infamous symbolism associated with criminal execution in Italy and France, where the anatomical figures were designed and published. In England, in the middle decades of the seventeenth century, a significant impetus not only for ecclesiastical and political reform, but also for a variety of programmes for universal reformation, was provided by millenarianism. Andrew Bradstock's study of Gerrard Winstanley presented in Chapter 6 considers the Digger theorist's debt to new ideas about husbandry and developments in agricultural practice. The essay places Winstanley within the early-modern scientific world view, informed as it is by belief in magic and alchemy, hermeticism and mysticism, in order to explicate Winstanley's economic and political programme and theological ideas. On Winstanley's indebtedness to mysticism, Bradstock does not come to a firm conclusion, but he does argue that Winstanley must be seen as a person of his time in terms of his millenarian convictions. In Chapter 7, Carola Scott-Luckens approaches the writings of Winstanley from a rather different angle, comparing the utopian and radical socialist aims of his community to the outwardly dissimilar practical reforms promoted under the auspices of European-born intellectual and correspondent Samuel Hartlib. Scott-Luckens surveys Digger ideals and husbandry manuals, before considering in detail publications produced by the Hartlib circle. Hartlib's own reform proposals attracted government funding, and represent, perhaps, the last attempt to promote knowledge and to utilise the nation's resources within a specifically ethical and Christian context.

Issues of gender and sexuality are the focus of Part III. Recent work in this field has emphasised the ways in which categories of sex as well as gender and sexuality were constructed in the early-modern period.[27] The section opens with Bronwen Price's re-evaluation of the works of the often marginalised figure of Margaret Cavendish (excluded as she was from membership of the Royal Society). Price argues here in Chapter 8 that Cavendish's writing makes important interventions with contemporary modes of knowing. She focuses primarily on the exploration of empiricism and rationalism in Cavendish's 1666 text, *The Blazing World*, which she characterises as having a protean quality. Price examines the way in which this text engages with some of the central ideas of Baconian and Cartesian epistemology, in order to recall and revise this; and suggests that Cavendish does this from a specifically feminine, perhaps even feminist, perspective. The subject of women's exclusion from the realm of natural philosophy and the field of science is continued in Elaine Hobby's Chapter 9, on Jane Sharp's *The Midwife's Book* (1671). This analysis is located within a broader history of both the midwifery manual and of midwifery itself; the latter a traditionally female territory which became increasingly invaded by men. Hobby concentrates on Sharp's use of and resistance to 'scientific' language in her manual, illustrating how her teasing refusal of the emerging medical discourse of seventeenth-century England reveals her analytical awareness of the gendered biases of masculine knowledge. However, her sometimes witty dismissal of male self-importance indicates that she was (erroneously) untroubled by the threat presented by the emergent man-midwives. In the following chapter, Ruth Gilbert looks at ways in which Renaissance men appropriated images of birth to describe both their textual productions and their intellectual, technological, and mystical aspirations and achievements. Drawing on both literary and scientific texts (including works by Milton, Bacon and William Harvey), she notes that male births range from images of pregnant men to the representation of hermaphroditic writing subjects, and focuses on how such images contribute to alchemical experiments and inform the masculinism of the new science. The last chapter in this section (Chapter 11) is also concerned with the representation of gender and sexuality. Here Bettina Mathes surveys the circulation of the figure of the nymph in art, literature, anatomy and medicine in the seventeenth century that eventually culminates in the construction of nymphomania, or 'excessive' female heterosexuality. That the nymph should come to stand for what was viewed as 'deviant' female heterosexuality is remarkable since, in Renaissance art and literature, nymphs signalled female homoeroticism. The heterosexualisation of the nymph was accompanied by the devaluation of the erotic significance of the clitoris and labiae in both art and medicine. In Mathes's view, the history of the term 'nymphomania' thus reveals that heterosexuality is the result of displaced homosexuality, enabled by the intense intercourse of art and medicine.

The final section of the book is concerned with the overlap between theories of natural knowledge and colonialist ideology. In Chapter 12 Andrew Hadfield addresses a key treatise promoting early English colonial ventures in the Americas by Thomas Harriot. This illustrated text is apparently motivated by the desire for scientific objectivity in cataloguing the life of the American Indians, yet at the same time it promotes not only colonisation, but specifically English colonisation. Hadfield concludes that this superficially accurate and impartial scientific approach to Indian society is riddled with contradictions, and that ultimately it challenges fixed notions of civility and savagery and shows up conflicts in contemporary notions of ethnography. Richard Sugg is also concerned in Chapter 13 with the analysis of a colonialist treatise: in this case, John Donne's 1622 sermon to the recently formed Virginia Company. This sermon is the author's most extended prose discussion of the New World, and Sugg reads it alongside Donne's secular and religious poetry, and alongside comparable contemporary texts. In this sermon, Donne soberly admonishes the colonists to win Christian converts immediately, and gold and commodities only in the fullness of time. Sugg takes the view that the tone is quite distinct from the exuberant opportunist fascination with the New World expressed in Donne's other works, and that his approach is rather different to that adopted by other writers of his time. Like Sugg, Mary Baine Campbell is also concerned with re-reading the less-well-known work of a central seventeenth-century figure, and with identifying original trends within it. In Chapter 14, Campbell interprets Johann Kepler's moon-voyage dream vision, *Somnium*, not only as astronomical and allegorical, but also as generically affiliated with New World exploration narratives. She suggests that Kepler corrects the hierarchical and threatening political geography implicit in these texts with a travel account of a new 'world' – the moon. The notion of an alternative world (rather than a 'new world' to be converted to Old World norms and connected to its economy) is stimulating politically. The 'Lunar Astronomy's' presentation of this alternative world, and its process of reorienting our perceptions of the solar system, taken together, suggest a path not taken by the new science of which Kepler has been so notable a hero.

We end with a caveat: in structuring this volume as we have, we invite readers to view early-modern 'texts' – fictional writings, poems, paintings, illustrations, treatises, prefaces, lectures and so on – from a very modern perspective, adopting a taxonomical system that makes sense in the twenty-first century. Categories or themes and topics such as philosophy, religion, politics, gender, sexuality, colonisation, imperialism, even science, reflect the world view of the contemporary academic and scholar but would not, or not necessarily, have been meaningful to seventeenth-century writers, artists and natural philosophers. Nonetheless, in putting together an interdisciplinary volume entitled *The Arts of 17th-Century Science*, we, the editors, hope that this collection of essays will continue the process of interrogating and dismantling the ahistorical and anachronistic opposition of the spiritual and the empirical, the rational and the imaginative, art and science.

Notes

1. Milton (1953), *Prolusions*, Exercise VII, in *Complete Prose Works*, 8 vols, ed. D.M. Wolfe, New Haven: Yale University Press, vol. 1, p. 296.
2. These aphorisms are taken from Bacon's *Advancement of Learning* (1605). See Bacon (1857–74), *Works of Francis Bacon*, 14 vols, ed. J. Spedding et al., London: Longman, vol. 5, p. 311. See Steadman, J. (1971), 'Beyond Hercules: Francis Bacon and the Scientist as Hero', *Studies in the Literary Imagination*, **4**, 3–47.
3. For further discussion of the connection between Milton's natural philosophy (in this case vitalism) and his political agenda, see Rogers, J. (1996), *The Matter of Revolution: Science, Poetry, and Politics in the Age of Milton*, Ithaca: Cornell University Press.
4. Recent scholarship about Britain's imperial ambitions in this period is vast, but for influential assessments see Helgerson, R. (1992), *Forms of Nationhood: The Elizabethan Writing of England*, Chicago: University of Chicago Press; Knapp, J. (1992), *An Empire Nowhere: England, America and Literature from Utopia to the Tempest*, Berkeley: University of California Press; Fuller, M. (1995), *Voyages in Print: English Travel to America, 1576–1624*, Cambridge: Cambridge University Press; Hadfield, A. (1998), *Literature, Travel and Colonial Writing in the English Renaissance 1545–1625*, Oxford: Clarendon; Linton, J.P. (1998), *The Romance of the New World: Gender and the Literary Formations of English Colonialism*, Cambridge: Cambridge University Press.
5. On the history of the Royal Society see Hunter, M. (1981), *Science and Society in Restoration England*, Cambridge: Cambridge University Press; see also Hunter, M. (1995), *Science and the Shape of Orthodoxy: Intellectual Change in Late Seventeenth-Century Britain*, Woodbridge: Boydell.
6. Rogers (1996), p. 1.
7. Svendsen, K. (1956), *Milton and Science*, Cambridge, MA: Harvard University Press; Fish, S. (1967), *Surprised by Sin: The Reader in Paradise Lost*, London: Macmillan.
8. Milton (1966), *Paradise Lost*, ix, 807–10, *Poetical Works*, ed. D. Bush, Oxford: Oxford University Press, p. 389.
9. Edwards, K. (1999), *Milton and the Natural World: Science and Poetry in Paradise Lost*, Cambridge: Cambridge University Press, p. 16.

10. See p. 174.
11. Milton (1966), *Paradise Lost*, ix, 202–3.
12. Bacon, F. (1857–74), *The Works*, ed. J. Spedding et al., 14 vols, London: Longman, vol. 3, p. 156.
13. Edwards (1999), p. 19.
14. Ibid., p. 21.
15. Milton (1966), *Paradise Lost*, ix, 276–7.
16. Mintz, S.I. (1952), 'The Duchess of Newcastle's visit to the Royal Society', *Journal of English and Germanic Philology*, **51**, 165–76.
17. Edwards (1999), pp. 19–21.
18. Porter, R. (1991), 'Introduction', in S. Pumfrey, Paolo L. Rossi and Maurice Slawinski (eds), *Science, Culture and Popular Belief in Renaissance Europe*, Manchester: Manchester University Press, p. 4. For an important examination of the relationship between court culture and the new philosophy, see Biagioli, M. (1993), *Galileo, Courtier: The Practice of Science in the Culture of Absolutism*, Chicago: University of Chicago Press.
19. See Linden, S.J. (1996), *Dark Hierogliphicks: Alchemy in English Literature from Chaucer to the Restoration*, Lexington, KT: University of Kentucky Press, pp. 1–36.
20. See Debus, A.G. (1978), *Man and Nature in the Renaissance*, Cambridge: Cambridge University Press, pp. 11–15.
21. Webster, C. (1975), *The Great Instauration: Science, Medicine and Reform 1626–1660*, London: Duckworth, p. xv.
22. On this term see Sawday, J. (1995), *The Body Emblazoned: Dissection and the Human Body in Renaissance Culture*, London and New York: Routledge, pp. 1–15.
23. Shapin, S. (1996), *The Scientific Revolution*, Chicago: University of Chicago Press, p. 3.
24. *Religio Medici*, in Browne, T. (1967), *The Prose of Sir Thomas Browne*, ed. N.J. Endicott, New York: Anchor Books, p. 47. We are grateful to Anthony Archdeacon for this reference.
25. Webster (1975), p. xv.
26. Shapin, S. and Schaffer, S. (1985), *Leviathan and the Air Pump: Hobbes, Boyle, and the Experimental Life*, Princeton: Princeton University Press; Albanese, D. (1996), *New Science, New World*, Durham, NC: Duke University Press; Daston, L. (1998), *Wonders and the Order of Nature, 1150–1750*, London: Zone Books. For a detailed bibliographic essay, see Shapin (1996), pp. 167–211.
27. See Laqueur, T. (1990), *Making Sex: Body and Gender from the Greeks to Freud*, Cambridge, MA: Harvard University Press. See also, for example, Fradenburg, L. and Freccero, C. (eds) (1996), *Premodern Sexualities*, New York: Routledge.

CHAPTER ONE

The Transparent Man and the King's Heart

Jonathan Sawday

I

'There was no such thing as the Scientific Revolution, and this is a book about it.'[1] So Steven Shapin begins his trenchant critique of the idea of a revolution in natural philosophy said to have taken place in Western Europe at some point during the sixteenth and seventeenth centuries. We may (or we may not) agree with those intellectual and cultural historians who reject the idea of a discrete 'revolution' in human understanding visible in the writings of natural philosophers of the sixteenth and seventeenth centuries. We may, indeed, choose to understand the period as (in Shapin's words) 'a diverse array of cultural practices aimed at understanding, explaining, and controlling the natural world' which owe as much to medieval patterns of thought and analysis as they do to anything which might be described as proto-modern.[2] This present short essay, however, is not concerned with the historical concept of a 'revolution' in science, one way or the other. Rather, I wish to focus on a single episode in which we can glimpse something of the *texture* of seventeenth-century modes of enquiry in operation.

My purpose in exploring that texture is twofold. First, I want to show how 'science' or 'natural philosophy' in the seventeenth century (as much as in the twenty-first) never operates in isolation from other cultural practices. Second, I should like to suggest how the language of scientists in the period is rarely anything other than, at best, opaque. In other words, the 'meaning' of a seventeenth-century scientific text may be as difficult to confirm as the 'meaning' of the more familiar (to literary historians) poetic and imaginative texts of the period. Nor should this surprise us: this was, after all, an age in which the poets and the natural philosophers shared the same education in rhetoric at the universities, read the same authors, and belonged to the same intellectual circles. The poet John Dryden, we should recall, was one of the founder members of the Royal Society, whilst Abraham Cowley sensed no embarrassment in publishing, in 1661, *A Proposition for the Advancement of*

Experimental Philosophy which was to establish the Royalist poet as one of the foremost apologists for the 'New Philosophy' of reason. Moreover, in the case of scientific texts, opacity is not simply a function (as traditional historians of science have tended to assume) of the relative primitivism of scientific technique, and nor is it the inevitable outcome of poor (by modern standards) instruments of observation and calibration available to the natural philosophers. In this respect, perhaps we have been blinded by the *contemporary* rhetoric of science to the less familiar rhetoric of the contemporaries of Donne or Milton.

Science in the early-modern period subsists within a larger cultural code, as did other forms of writing with which we are often better acquainted. This code may be determined by political events, belief-systems, even unconscious fears and desires, which would seem to have, at first sight, little to do with the austere communion with nature recommended by Bacon and his successors. Hence, my endeavour in this essay is to try and read a seventeenth-century scientific 'episode' as it might have been read and interpreted by the contemporaries of Andrew Marvell or Oliver Cromwell, rather than by the contemporaries of Stephen Hawking or Richard Dawkins.

II

To observe the beating human heart in a subject who feels no discomfort, let alone pain, is, in the twenty-first century, a commonplace of medical science. Such sights have become the stuff of TV documentary and entertainment. But, prior to the mid-nineteenth century, when the use of anaesthesia in surgical operations was established, such a sight would have been beyond rarity.[3] Nevertheless, the prospect of a living human interior was afforded two individuals in the seventeenth century, and the record of their observations offers a possibly unique insight into what I have termed above the texture of seventeenth-century science.

Some time before 1640, King Charles I learned of the curious condition of a young Irish nobleman, Hugh Montgomery, third Viscount of the Ards, and later first Earl of Mount Alexander. In his childhood Montgomery had suffered an injury to his left side, which had produced an extensive abscess. This abscess, which had never properly healed, was protected by a metal plate which, when removed, granted, quite literally, a window into the human interior. His curiosity aroused, the King ordered his physician, William Harvey, to examine Montgomery. Harvey takes up the story:

> … having … declared unto [Montgomery] the Cause of my Visit, by the King's Command, he discovered all to me, and opened the void part of his left side, taking off that small plate, which he wore to defend it against any blow or

> outward injury. Where I presently beheld a vast hole in his breast, into which I could easily put my three Fore-fingers and my Thumb: and at the first entrance I perceived a certain fleshy part sticking out, which was driven in and out by a reciprocal motion, whereupon I gently handled it in my hand. Being now amazed at the novelty of the thing, I searched it again and again ...[4]

In reading Harvey's account of his inspection of Montgomery, we may be immediately struck by the apparent terseness of the description; its determination to record only what can be seen. It is as if Harvey had already anticipated Thomas Sprat's famous 1667 dictate upon scientific language and the language of scientists, that they should 'separate the knowledge of *Nature*, from the colours of *Rhetorick*, the devices of *Fancy*, or the delightful deceit of *Fables*'.[5]

It is Harvey's commitment to observation, indeed, which allows him, sceptically, to question what it is that he is actually examining within the young Irish nobleman's interior. The structure which Harvey had so eagerly reached after in the young man's interior proved to be not, as others had supposed, the lungs palpitating with respiration, but something altogether more vital. What Harvey knows he is seeing is a living, beating, human heart. Aware that no written account could do justice to the sensational nature of what he had not only observed, but handled, Harvey determined on bringing Montgomery to the King:

> ... instead of an Account of the Businesse, I brought the Young Gentleman himself to our late King, that he might see, and handle this strange and singular Accident with his own Senses; namely, the Heart and its Ventricles in their pulsation, in a young and sprightly Gentleman, without offense to him: Whereupon the King himself consented with me, That the Heart is deprived of the Sense of Feeling. For the party perceived not that we touched him at all, but meerly by seeing us, or by the sensation of the outward skin.[6]

Why would 'an Account of the Businesse' not have sufficed the curiosity of the King? Harvey does not explain, but the inference we might draw is that what has been glimpsed within the young man's interior is so sensational that only by seeing it for oneself will full justice be done to the phenomenon. In other words, the King must replicate, alongside his physician, the very experience of reaching within Montgomery's frame. The implication of the description is that the King himself, along with Harvey, thrust his fingers into Montgomery's side. But there is room for some ambiguity here. Did Charles actually explore the 'void' as Harvey had done? That seems to have been Harvey's intention in bringing the young man into the King's presence, but all we learn is that the King 'consented with me'.

In the ambiguity of the encounter between Montgomery, the physician and the King, we might be reminded of another famous encounter with the human interior: that recorded in the scriptures when Christ appeared amongst the disciples following his resurrection. It is as if the Gospel account of 'Doubting Thomas' (itself the subject of one of the more sensational, and better-known, examples of Baroque art: Caravaggio's *Incredulity of St Thomas* of 1603) had been brought to life in seventeenth century London.[7] Indeed, the Gospel story is easily transposed onto the tableau of monarch, scientist and young Irishman: 'Reach hither thy hand, and thrust it into my side: and be not faithless but believing' (John 20:27). Except that the truth of what was being revealed in Whitehall was not a matter of religious faith, but of scientific certainty: '… the Heart is deprived of the Sense of Feeling'.

Or were things so certain? In modern terms, to observe that the heart is 'deprived of the sense of feeling' is to say little more than that it does not possess a nervous apparatus capable (like the skin) of transmitting certain kinds of sensation to the brain. The heart, in other words, does not touch, smell, taste, see or hear. Instead, its 'office' (to use the seventeenth-century term) is to act as the motor or pump of the dynamic circulatory system. But this was not how the heart was understood in the earlier seventeenth century. 'By the early modern period' Robert Erickson has observed 'the word "heart" had come to mean a variety of things':

> … the centre of all vital functions, the source of one's inmost thoughts and secret feelings or one's inmost being, the seat of courage and the emotions generally, the essential, innermost, or central part of anything, the source of desire, volition, truth, understanding, intellect, ethics, spirit. *It was the single most important word referring both to the body and the mind.* No other word performed what 'heart' did, and no other word today quite replaces it [author's emphasis].[8]

In other words, to say that the heart of the young man whom Harvey and the King so busily explored – and that they perhaps handled – was 'deprived of the sense of feeling' was to make rather more than a strictly physiological observation. But what, other than physiology, might have been at stake? Carefully timing the 'pulse and the different rhythme' of the beating substance which they were inspecting, and then comparing that rhythm to Montgomery's pulse in his wrist, Harvey and the King were in no doubt as to the nature of what it was that they were witnessing. However, by passing through the body's hitherto impermeable surface barrier, they were able to attest to a larger philosophical and linguistic phenomenon: the heart could not think or feel. It was not (as Aristotle had once claimed) the seat of sensation, courage, honour, fortitude, generosity, kindliness, love, cruelty, anger or indifference. Neither was it (as Edmund Spenser had described it in 1590 in Book II of *The Faerie Queene*) the 'goodly parlour' wherein were held to reside the affections and the seat of the sensible (as opposed to rational) soul.[9] Instead, the heart is witnessed as an essentially Cartesian mechanism.

Yet, there is an irony in the image of the King and the physician groping and peering into the living human interior which needs to be remarked upon. In 1628, a dozen or so years before his encounter with the Transparent Man, Harvey had published the work which was to establish his reputation throughout Europe: *Exercitatio anatomica de motu cordis et sanguinis in animalibus* ['An Anatomical Exercise concerning the movement of the heart and blood in Animals']. Famously, *De motu cordis* established, through a set of rigorous observations and empirical tests, the doctrine of blood circulation within the animal interior. Yet, as has often been observed by medical historians, Harvey was something of a reluctant revolutionary. In political terms, indeed, his royalism and Anglicanism set him apart from many of his contemporaries who, in the mid-seventeenth century, were to embark on a radical experiment in constitutional and political change.

Harvey's conservatism is nowhere more apparent than in the dedicatory words with which he prefaced *De motu cordis* in its 1628 (Latin) version. Addressing the King, Harvey wrote of his work that he hoped it would strike a particular chord with the monarch since it was concerned with the heart which mirrored, in the human and animal body, the constitutional role of the King within the commonwealth. For the King is (Harvey wrote) 'the heart of the state; from him all power arises and all grace stems'. And he continues:

> An understanding of his heart is thus of service to the king as being a very special portrayal, if on a more modest level, of his own functioning. Placed, best of Kings, as you are at the summit of human affairs, you will at least be able to contemplate simultaneously both the central organ of man's body and the likeness of your own royal power. Accept, therefore ... this new account of the heart. For to you, who are yourself the new splendour of this age, and indeed its whole heart, its central figure abounding in virtue and grace, we rightly refer whatever good obtains in this England of ours, whatever pleasure in our life within it.[10]

Reading this fulsome and entirely conventional dedication, and comparing it to the later description of the encounter with an actual living heart in the body of young Hugh Montgomery, we may begin to sense at least the potential for a moment of embarrassment. In the 1628 dedication to his King, Harvey links the constitutional part played by the monarch within the state to the physiological functioning of the heart within the body in terms of power, virtue and grace. The crucial and (to the modern reader) unfamiliar term in this triptych of royal and cardiac function is the term which is repeated: grace. In claiming that both Charles and the heart are the seat of 'grace' (*gratia*) Harvey is referring the reader back to the medieval idea of the king as *christomimetes* – an actor or impersonator of Christ. This act of impersonation is by no means, as we moderns might expect, blasphemous; rather, it is a matter of equivalence between Christ and the King. Ordained by God, and anointed by the Grace of God, the King's power within the state is held to be analogous to that held by the anointed Old Testament kings of Israel, who themselves were believed to foreshadow the advent of the incarnate Christ. This is the doctrine of the 'King's Two Bodies' whereby the monarch was held to be (in the famous account of Ernst Kantorowicz) a 'twinned' being, partaking (as did Christ) in both the human and the divine. Or as an anonymous early twelfth-century Norman manuscript, known to the Elizabethan Archbishop Matthew Parker, put the matter:

> We thus have to recognize [in the king] a twin person, one descending from nature, the other from grace … . One through which, by the condition of nature, he conformed with other men: another through which, by the eminence of [his] deification and by the power of the sacrament [of consecration], he excelled all others. Concerning one personality, he was, by nature, an individual man: concerning his other personality, he was, by grace, a Christus, that is, a God-man.[11]

'Power', 'virtue' and 'grace' – terms which link together the King in his commonwealth with the heart in the body – refer us to the explicit doctrine of 'divine right' which Charles's father, James VI and I, had done so much to uphold, and which Charles himself was so stubbornly (and in the end disastrously) to maintain. In essence, what Harvey is remarking upon (perhaps even encouraging the King to believe) is that his role in the state is one of nutriment, in both an economic and spiritual sense. By reading *De motu cordis*, Harvey implies, the King will learn that his political function is now to be understood as not merely part of the natural order, but in conformity with the most rigorous observation of an emergent empirical science.[12]

In 1628, of course, it was perfectly possible to maintain such a link between the natural body and the body-politic. To say of the King that he was 'like the heart' was to utter no more than a commonplace of medieval and Renaissance political theory. But the account of Harvey and the King peering into an *actual* human body to observe a living heart performing its office within the human interior takes on a new resonance when it is recalled that Harvey's own record of the observation of Hugh Montgomery was written after 1649 – after, that is, the 'Royal Actor' (as Marvell was to describe Charles I) had met his fate on the scaffold in Whitehall.[13] For it was as an act of memorialisation of the executed monarch's scientific curiosity, as much as a record of an unusual phenomenon, that Harvey, in 1653, had published this story, in which the King appears as 'our late King Charles'. In this respect, perhaps we should be reading the account of Charles's encounter with the living heart of one of his subjects as a more directly political expression on the part of Harvey. Perhaps, indeed, this digressive interlude should be read alongside a text such as the explosively popular *Eikon Basilike* of 1649, where Charles made his first appearance as the 'royal martyr' (see Figure 1.1).[14]

Hugh Montgomery seems to have led a full and eventful life, virtually

1.1 Charles I, King of England, *Eikon Basilike*, London, 1649, frontispiece.

unaffected by what might have been thought to have been (in any age) a potentially disabling condition. But he could not lead his life entirely unassisted, for his wound continued to produce what Harvey termed an 'excrescent funguous substance (as is usual in foul ulcers)' which required continual assistance from Montgomery's manservant:

> The Young Gentleman's Man did by dayly warm injections deliver that fleshly accretion from the filth and pollutions which grew about it, and so clapt on the Plate; which was no sooner done, but his Master was well, and ready for any journey or exercise, living a pleasant and secure life.[15]

At the time of writing this account, some time between the execution of the King in 1649 and 1651 (when the text was first published in Latin), Harvey would have known of the fate of many of the supporters of the King as they entered what the Royalist poet Richard Lovelace was to term the 'cold Time and frozen Fate' of the Republic. It is not impossible that he was aware of the subsequent career of the young man whose interior he and the King had once scrutinised with such fascination. Montgomery had played a prominent part in supporting the King's cause in Ireland, commanding the Royalist army which, after May 1649, had seized, successively, Belfast, Antrim and Carrickfergus. But, eventually, Montgomery was forced to surrender to Cromwell, and was brought back to London. Banished, at first, to Holland, he was eventually allowed to return to Ireland where he led a life which was severely circumscribed, enduring at least one period of imprisonment in Kilkenny Castle. Whether or not Harvey knew the details of this story at the time he wrote the account of the inspection of Montgomery, he would have known that the prospects for the executed King's former supporters (particularly one as prominent as the young Irishman) were, to say the least, gloomy. Harvey's audience, too, would have known of the fate of those such as Montgomery who had championed the Royalist cause. And, of course, they would have been all too aware not only of Charles's eventual fate, but of the hagiographic whirlpool which came to surround his memory. So, to ask whether the text could have been read as a form of political allegory by the contemporaries of Harvey is, I would argue, to ask the wrong question. Rather, we might ask how possible it would have been *not* to read the story of the Transparent Man as a narrative suffused with ironic political overtones.

The story of Harvey's remembrance of his encounter with the Transparent Man in the King's presence is something more than a simple account of an unusual medical condition. Reading with hindsight (and mindful that Harvey wrote the story in the full knowledge not only of the King's own fate, but also of that of so many of his supporters after 1649), what begins to strike the modern reader as much as it must have struck Harvey's contemporaries is how richly allusive the story is as a political allegory. Peering into the human body, we encounter an organ (the heart) disrobed of its former majesty, unfeeling and blindly pumping. The dense web of allegory, resemblance and metaphor which had surrounded the heart in Galenic and Aristotelian physiology has collapsed as completely as the quasi-divinity which had surrounded the King's political *persona* prior to 1649; moreover, the formerly majestic heart is now surrounded by a 'fleshly accretion from the filth and pollutions which grew about it'. What, after 1651, has the heart become? Certainly, the political body is still functioning, but its core – the beating heart, which had once been the seat of grace – is now transformed. 'The King himself consented with me' Harvey had written, recalling his inspection of Montgomery, 'That the Heart is deprived of the sense of feeling'. We should be alert to the ambiguity of that observation. For in writing that the heart is now 'deprived of the Sense of Feeling' Harvey manages to suggest not that it *never* possessed feeling, but that it *no longer* can be treated as an organ of sensation within the economy of the animal. It has become, as the King had also become by the time this passage was written, an organ from which (in the words of the 1628 dedication to *De motu cordis*) 'grace', 'power' and 'virtue' can no longer flow. The heart, in other words, had been usurped. The animal heart lives on, it is true; but the beating heart of the state is stilled. And hence my conclusion: what Harvey, consciously or unconsciously has sought to evoke is not the living heart of a young Irishman, but the stilled heart of the King who once stood by his side and, in yet another rehearsal of a scriptural trope, witnessed an emblematic representation of his own authority and power.

Notes

1. Shapin, S. (1996), *The Scientific Revolution*, Chicago and London: University of Chicago Press, p. 1.
2. Ibid., p. 3.
3. The key date in the use of anaesthesia is usually taken to be 1853, when Dr John Snow administered chloroform to Queen Victoria at the birth of Prince Leopold. See Porter, R. (1997), *The Greatest Benefit to Mankind: A Medical History of Humanity from Antiquity to the Present*, London: Harpercollins, p. 367.
4. Harvey, W. (1653), *Anatomical Exercitations concerning the generation of living creatures*, London, p. 286. For an account of the story, see Keynes, Sir Geoffrey, (1978), *The Life of William Harvey*, Oxford: Clarendon Press, pp. 155–7. The story forms the basis for one of the most dramatic episodes in Rose Tremaine's novel *Restoration* (1989).
5. Sprat, T. ([1667] 1966), *The History of the Royal Society of London*, ed. J.I. Cope and H.W. Jones, St Louis and London: Washington University Press/Routledge and Kegan Paul, p. 62.
6. Harvey (1653), p. 287.

7. For a discussion of this painting see Sawday, J. (1997), 'Self and Selfhood in the Seventeenth Century', in Porter, R. (ed.), *Rewriting the Self: Histories from the Renaissance to the Present*, London and New York: Routledge, pp. 32–8.
8. Erickson, R.A. (1997), *The Language of the Heart 1600–1750*, Philadelphia: University of Pennsylvania Press, p. 11. I am indebted to Erickson's richly suggestive account of Harvey and the language of the heart in the present essay.
9. Spenser, E. (1977), *The Faerie Queene*, ed. A.C. Hamilton, London and New York: Longman, p. 253.
10. Harvey, W. (1990), 'Movement of the Heart and Blood in Animals', trans. K. Franklin, intro. Wear. A., *The Circulation of the Blood and other Writings*, London: Everyman, p. 3. This is a modern English translation of the Latin text of 1628.
11. The passage from the 'Norman Anonymous' is cited in Kantorowicz, E.H. (1957), *The King's Two Bodies: A Study in Medieval Political Theology*, Princeton, NJ: Princeton University Press, p. 48.
12. It perhaps should be recalled here that 'grace' (*gratia*) in the form of a crown of thorns forms a significant feature of the frontispiece of *Eikon Basilike* (1649). On what has been termed 'the literary dimensions of [Harvey's] political philosophy' see Rogers, J. (1996), *The Matter of Revolution: Science, Poetry and Politics in the Age of Milton*, Ithaca: Cornell University Press, pp. 16–27.
13. It is difficult to resist the suspicion that it was the idea of the King as *christomimetes* which Andrew Marvell had in mind when he described Charles I on the execution scaffold as the '*Royal Actor*' in 'An Horatian Ode upon Cromwell's Return from Ireland' of 1650. See Margoliouth, H.M. (ed.) (1971), *The Poems and Letters of Andrew Marvell*, 2 vols, Oxford: Clarendon Press, vol. I, p. 92.
14. The popularity of *Eikon Basilike* is affirmed when we recall that the text ran through 35 editions within the first year of its appearance.
15. Harvey (1653), p. 286.

PART I
PHILOSOPHY, THOUGHT AND NATURAL KNOWLEDGE

CHAPTER TWO

'Things Which Are Not': Poetic and Scientific Attitudes to Non-entities in the Seventeenth Century

Anthony R. Archdeacon

Filling the void: theory and evidence

According to Hobbes's *Leviathan* (1651), the word 'body' signified

> that which filleth, or occupyeth some certain room, or imagined place; and dependeth not on the imagination, but is a real part of what we call the Universe. For the Universe, being the Aggregate of all Bodies, there is no real part thereof which is not also body; nor anything properly a body, that is not also part of (that aggregate of all bodies) the Universe.[1]

Interestingly, Hobbes's materialist account of the universe left little room either for products of the imagination – fictions – or for a notion revolutionised in his lifetime – empty space. The justification for considering these two phenomena together is exactly the fact that, in the seventeenth century, for the first time, one might have treated both as phenomena. The 'idea' was to become reified in Cartesian discourse, whilst the vacuum was to become an object of experimental observation. This essay considers how poetic and scientific discourses responded to these changes in perceptions of, and attitudes to, things. The vacuum had long been excluded from the category of 'things': it was a nothing, a concept for which people did not bother to seek a correlative in actuality. If not anathema, it was at least illogical, barely even imaginable. What would a vacuum look like? At the beginning of the seventeenth century, the traditional way of representing the void pictorially, blackness, was about to become an obsolete metaphor, as the status of the term 'vacuum' shifted from the ideal to the phenomenal. By 1660 there was no more need for metaphor or abstraction: one could look at the diagrams of Robert Boyle's air pumps in his publication, *New Experiments Physico-Mechanical, Touching the spring of the Air and its effects*.[2] This shift was made possible by new discoveries in natural knowledge. The question addressed in this essay is whether such movements in the discourse of natural philosophy were seen, conversely, in literary/poetic discourse, where metaphor was the dominant conventional mode. Was poetry supplanted, or at least demoted, by the ascendancy of a modern scientific discourse? If so, then the shift in perceptions, or conceptions, of the vacuum seems to provide a dramatisation of this discursive movement.

The intersection of the discursive currents concerned with fictions and with vacuums is cognate with the tension between speculative and practical mathematics described by Jess Edwards in this volume (see Chapter 3). Edwards points out that the language of mathematics was itself metaphorical, as exemplified by the mathematical point. An actual dot on the page stands in, metaphorically, for something more real – the mathematical idea of a point. Because the mathematical point cannot exist except as an idea, then the whole of mathematics had seemed to be an ideal construct, thereby encouraging mystical, hermetic interpretations of the world. Pierre de la Ramée had promoted a more pragmatic approach, emphasising the use rather than the 'nature' of the mathematical figure. Before the seventeenth century the vacuum was, like the mathematical point, a 'nothing' in philosophical terms – impossible and unimaginable. Robert Kaplan has described the transition between the use of a dot and the use of a hollow circular zero for the mathematical cipher, and the ontological, even cultural, implications of both.[3] This essay suggests that there was something similar about the impact of the 'actualisation' of the idea of vacuum.

Hobbes shared with most seventeenth-century thinkers the conviction, notwithstanding new experimental evidence to the contrary, that the existence of a vacuum was an impossibility – indeed, something of a self-contradiction. From Parmenides through the Middle Ages (and in spite of various inconclusive attempts to demonstrate the force of a vacuum by Adelard of Bath, Alexander Neckham and, most famously, Roger Bacon), the existence of a vacuum in nature had remained at the speculative level. It seems that, even during the seventeenth century, empirical criteria tended to be jettisoned in an attempt to explain, by a priori assumptions such as the existence of the ether, the phenomena observed by experimental scientists.[4] It is notable that in the passage quoted above Hobbes referred twice to imagination in relation to the 'place', or geometric space, occupied by body. Space was often called 'imaginary' in medieval scholastic terminology, because of its abstractness. In sixteenth-century metaphysics, it was thought of as an emptiness which bodies filled – a sort of shadowy negative of body, but still real. Hobbes, however, goes beyond a mere contrast of body and space, counterposing space and reality: space is only imagined, body is real.

The implications of stating that space is merely a convenient geometrical fiction went beyond the challenge to the Cartesian identification of body with extension. Hobbes was equally dismissive, in the same chapter of *Leviathan*, of the idea of 'spirit' or, as he tendentiously put it, 'incorporeal substance'. Since substance *is* body, the term makes no more sense than 'incorporeal body'. Hobbesian materialism denied reality to the spiritual world and the imaginary world at once, with the implication that the spiritual was imaginary. Both concepts are described as 'idols of the brain, which represent Bodies to us, where there are not, as in a looking-glasse, in a Dream, or to a distempered brain waking'.[5] These notions, which proceed from disorderly tumult in the brain, would have included the idea of vacuum. And yet, six years before the publication of *Leviathan*, the 'London group' of experimental scientists – the forerunners of the Royal Society – were studying Evangelista Torricelli's barometric experiment, which had appeared to demonstrate the possibility of producing a vacuum.[6] To see how these two items are related, one needs to consider the reception accorded to vacuum experiments in the mid-seventeenth century.

The excitement amongst scientists all over Europe at the new experimental demonstrations of vacuums had come about by chance. Torricelli's experiment – made first with a column of water in 1643, then a column of mercury in 1644 – had been devised to demonstrate the weight of air, but the controversial by-product of this was to suggest that weight beyond one atmosphere, or 30 inches of mercury, could indeed evacuate the space at the top of a sealed tube.[7] The level of mercury in the tube, inverted in a dish of mercury, would therefore rise or fall according to air pressure. No practical use for this was explored by Torricelli, who died suddenly in 1647, but barometric measurements, and even weather forecasting, were subsequently carried out by Otto von Guericke. Guericke himself became famous for his invention of the air pump (1650), and for the spectacular demonstration of his 'Magdeburg Hemispheres' in 1654. When air was drawn out through a valve by means of his air pump, Guericke found that air pressure would counter the most extreme attempts to separate the two halves of a metal sphere.[8]

Guericke had been sure that he was producing at least a partial vacuum, but, despite experimentation using vacuum pumps carried out by Boyle, Newton and others in England, the issue of what, if anything, was inside evacuated vessels remained as contentious as ever.[9] There was a bewildering variety of conclusions issuing from, on the one hand, the experimental evidence and, on the other, the dictates of theology and/or 'reason'. The apparent production of 'terrestrial' vacuums contradicted both Aristotelian thought and Christian theology: not only was the void normally described as that which was outside and/or before the created universe, it was also almost a synonym of 'nothing'. In the words of the essayist Edward Daunce in the 1580s, a vacuum was considered 'the natural residence of nothing'.[10] For this reason, some considered that to posit the existence of a vacuum was contrary to God who had banished nothingness by his creation – an Augustinian stance taken, for example, by Thomas Erastus (1523–83).[11] Theological wranglings over the status of cosmic voids continued throughout the sixteenth century, from the Scottish John Major (1467/9–1550) to the Portuguese Jesuit Pedro da Fonseca (1528–99).[12]

The emphasis upon empirical investigation rather than speculation in the seventeenth century did little to clarify the picture with regard to vacuums, which was all the more obscure because in practice such discursive distinctions were not employed by experimenters themselves. Torricelli had anxiously avoided getting involved in the philosophical and theological debates his work provoked, whilst Guericke himself used the doctrine of creation *ex nihilo* to support his argument that space was an uncreated nothing.[13] In France, where in 1644 Torricelli's barometric experiment aroused great interest amongst scientists, Blaise Pascal affirmed the existence of vacuums in barometers, citing the variable barometric readings at different heights as proof that the weight of air caused the phenomenon.[14] Pascal's writings were to dismiss 'l'horreur du vide', nature's abhorrence of the vacuum, and stirred up considerable philosophical as well as scientific discussion. It is difficult today to appreciate the conceptual shift involved in acknowledging the real existence of terrestrial vacuum. Pascal's Jesuit opponent Père Noël urged him to use the term 'éspace immaginaire' rather than 'le vide', to eschew some of the threatening implications for Christian doctrine of the existence of a void in nature.[15] Guericke had, after all, begun from Copernican premises, assuming only the possibility of empty space beyond the terrestrial world, across which magnetic action controlled the movement of the planets. Yet the sheer physical reality of his vacuum pump and iron hemispheres had already brought the vacuum out of the abstract realm of cosmological speculation, and into one as mundane as today's thermos flask.

Non-entities, or convenient fictions

The vacuum/nothing question might be taken as a paradigmatic example of that seventeenth-century polarisation of approaches to ontology usually characterised in terms of material–spiritual, or physical–metaphysical oppositions. Poetry had some role, through the notion of poetic fiction, in negotiating these ontological confusions as empirical science increased its intellectual influence during the century. To understand this requires consideration of the differences between seventeenth-century and the most recent perspectives on science, on poetry, and on the relationship between the two. The most obvious point to be made is that, unlike twentieth-century scientific discourse, Renaissance scientific language could be poetic. The classical precedent of Lucretius' *de Natura Rerum*, a particularly popular text at the turn of the seventeenth century, might be taken as a paradigm. In spite of Royal Society pretensions to a less poetic, plainer scientific language, there was no clear discursive distinction to be made. Alchemic texts, indeed, continued to favour poetic expression, along with elaborate visual symbolism, well into the century, along with elaborate visual symbolism. To take a different perspective, many Renaissance poets thought of themselves as employing a discourse which was inclusive, synthetic. This was the fantasy of humanistic poetry, involving a conception of the 'humanities' as an all-embracing study, with the formal structures of Renaissance verse providing a shaping, unifying expression to encyclopaedic knowledge. The development of a modern, exclusive scientific discourse was bound, therefore, to conflict with such poetic aspirations, yet just as predictably there were efforts to heal the emerging breach, to retain the role of metaphor in writing about the world – in a sense to preserve the dominance of a poetic discourse which had prevailed during the Renaissance.

The role of metaphor in negotiating new categories of being can be seen in the way, in the late-twentieth century, we adapted to the phenomenon of 'virtual' space. The virtual realm of the computer is enmeshed with the concrete realm. Physical and computer-generated worlds run together if not seamlessly, then with a seam of which we are increasingly oblivious. *Hyper-reality* has become virtually omnipresent, in both work and leisure environments, making the notion of fiction almost redundant, in the absence of any correlative *reality* to act as foil. Particularly interesting is that we rapidly found that the only way to orient ourselves in 'hyperspace' was to extend rigorously the spatial metaphor, with 'frames' and 'boxes', and of course 'windows', enclosing that space. There is no sense now that the metaphorical status of such virtual structures makes them less real: reality is now user-led, and the user wants a helpfully structured, easily navigable virtual world. Moreover, the user expects that world to be populated by objects, not programming scripts or DOS commands, therefore our use of computers has become almost exclusively 'object-oriented'. We click on an icon, a button or a banner, not even thinking what processes are 'really' taking place, because we are quite happy with an illusion so long as it is internally consistent. To say this is to say much the same as Petrus Ramus was saying about Geometry, and indeed Logic, in the sixteenth century: its content needed to be useful, and well ordered, even if its order was a product of imagination rather than the nature of things. The cultural shifts of the Renaissance and late twentieth century both disrupt the relation between reality, things and fictions.[16] Absolute opposition of fictionality and reality is not possible once we talk of metaphors as objects, of fictions as things, since in a basic sense reality means 'thingness'.

There might not have been quite so fluid a notion of reality in the seventeenth century as there is now. It is evident that the perspective of seventeenth-century thinkers was heavily reliant on an opposition of reality to image – both the image on the canvas or the page, and the image in the mind. Thence arose literary tropes which opposed *appearance* to reality, with art considered as mimetic. But, of course, reality was never understood in the simple way suggested by Baudrillard's schematic account of the historic phases towards the 'simulacrum'.[17] Reality was a highly contested area in the early-modern period, and particularly in the late-sixteenth century, when the (virtual) world of the stage started to push beyond its traditionally emblematic function towards representation. Shakespeare's sense of metadrama continually forced the issue: the 'wooden O' of the Globe theatre was both nothing and the whole world. The question then, as now, was what, if not things, are fictions? Non-things? Non-entities?[18] The conundrum of what is a non-entity leads us not only towards the logical problem of fictionality, and conveniently to the quotation from John Donne in my essay title. Donne was writing at a time when poets and other artists were reflecting upon the reality-status of their own fictions, and the relation of their fictions to the real (that is, non-fictional) world. The nature of that relation, which has been much discussed in recent years, was certainly less interesting to seventeenth-century thinkers than the problem of the distinction between real and unreal itself. Donne's phrase 'things which are not' illustrates his own preoccupation with that abstract enquiry recurrent in early-modern culture, beyond the ontological absolutes of Christian theology.

Renaissance writers – such as Shakespeare, John Donne, Margaret Cavendish and Thomas Browne – were directly concerned in their works with what *is*, in quite an abstract way. Metaphor was for them a way of addressing ontological uncertainty. There they contrast with the cartographers, the post-Baconian discoverers, the anatomists and the members of the Royal Society, whose assuredness about what *is*, and therefore about what things were proper objects of knowledge, defined their status as 'scientific' in the modern sense. The questioning of the category of existence had belonged, in the Middle Ages, to metaphysics and dialectics, but notwithstanding rearguard efforts by Counter-Reformation thinkers in the early seventeenth century, these were increasingly marginalised disciplines. The general and academic influence of Aristotelianism, and therefore of Metaphysics as a discipline, was declining by the beginning of the seventeenth century. Though it retained its status as a higher discipline within academia, in the new extra-academic media of intellectual exchange metaphysical discourse was being superseded by a new 'scientific' discourse. The establishment of the Royal Society signified a new kind of intellectual academy exactly because it was new; because it was ostensibly unencumbered by scholastic or other traditions, which had typically focused attention on abstract questions of essence and existence.

These were arcane, and of course Latin, traditions. Treatises in the vernacular on human knowledge and understanding from Descartes and Locke were reaching not just an academic fellowship, but a public readership too. Those works affected a reassuring certainty about what is and what is not: Descartes declared that simple ideas are irrefutably true; Locke that we can have clear and distinct ideas derived from either the senses or reflection. Their philosophical methodologies effectively stepped back from ontological questions, in the belief that our own ideas, passions and perceptions could be treated as data – as things in the mind, or mental objects. And the prose style of Descartes, mixing method with autobiography, crossed the boundary between the theoretical and the actual, so that he could talk about ideas as things in a way no one had before.

All in the mind? Fictions, from metaphysics to metaphor

Although the Cartesian way of talking about ideas was distinctively new, talk of 'things in the mind' was not a seventeenth-century innovation. Reference to *ens rationis*, which might translate variously as 'being of reason' or perhaps 'thing of the mind', can be found in the thirteenth-century *Summa Theologiae* of Thomas Aquinas. Following Suarez in his *Disputationes Metaphysicae* (1597), the phrase gained common currency in academic, metaphysical terminology as the name for fiction: things which existed *only* in the mind, and not in the real world.[19] Strictly speaking, *ens rationis* as used by Suarez was the name of a category of being, but semantic slippage could allow it to be the names for fictions themselves.

The relevance of this to poetic discourse is greater than one might expect. In the academic institutions of the seventeenth century, metaphysics was becoming increasingly the exclusive domain of theologians, particularly Jesuits from the Iberian peninsula like Suarez. Meanwhile, poets of the same period appear to have latched on to a way of speaking about the world which was losing general currency thanks to the spread both of scepticism and of the scientific projects which emerged in its wake. This perhaps points rather to metaphysical textbooks having been an available storehouse of various metaphors and abstractions about the world than to some intellectual movement towards metaphysics amongst poets. These books supplied educated poets with a vocabulary which translated into themes that signified 'elevated' poetry: soul, spirit, the absolute and the infinite – and, of course, that most abstract and fundamental opposition, being and not-being. At the turn of the century, John Donne's use of metaphysical language in his poetry drew, like many of the university-educated poets of his generation, on the recondite discourses of the schoolroom. It was a shared, exclusive (and male) language, drawing on the combined attractions of nostalgia and intellectual clubbishness, wittily playing with the metaphysical and dialectical ideas which had been the staple of their university years. This was their literary *materia prima*, which they endlessly formed and re-formed into sonnets and epigrams to amuse their peers and impress the rest.

Sir John Davies of Hereford, one of the less prominent metaphysical poets, was apt to draw on a wide range of philosophical and theological sources, as in this sonnet:

> The Stoicks, in their strange Philosophie,
> Make All, and Nothing, nothing but all one:
> Who say that this World Is, but yet deny
> That it hath any Essence of the owne.
> But, in our loves (deere Love) the same is true:
> For thou, being All, art mine, that nothing am?
> I nothing am that is not All thy due,
> So, All and Nothing's nothing but the same!
> Then, sith my Nothing and thy All all's one,
> Thou, All, I, Nothing, make an Unity:
> For, All to Nothing hath conversion,
> And Nothing, unto All, by sympathie:
> Then neede I (nothing) Thee (All) nothing feare
> But All, and Nothing still shall One appeare.[20]

Thematically and poetically, there are echoes of Donne's alchemy-inspired 'A nocturnall upon S. Lucies day', though the philosophical references are very different – Davies was alluding to the Stoic paradox of the unity of opposites, derived from Heraclitus. Numerous poems by Davies from the first two decades of the seventeenth century address the issue of 'thingness' and nothingness via theological and philosophical issues – whether evil, thought, mortal life, the material world, man, woman, love or knowledge are nothing. The epigrammatic style of this poem mimicked the sophisms of scholastic teaching in its deliberate paradox:

> … is ill nought? Why then it IS, though nought:
> But, Nought is nothing: then IS nothing? No.
> Yet it is nought, descending still from Ought
> So, then it IS, and yet, it is not so.[21]

'Nothing' became established as a favourite theme for epigrammatists: John Heywood, Sir John Harington, John Owen, Robert Hayman, Henry Parrot, John Heath, Robert Watkins and John Vicars all wrote epigrams on nothing. Most of these epigrams involved elaborate wordplay, recalling the logical conundrums which they would have had to solve as undergraduates. They seem to have been fascinated by the semantic and metaphoric possibilities of 'nothing', and this was often combined with the topic of penitential self-abnegation which made all mortals nothings, as in this epigram from Francis Quarles (1592–1644):

By nature, Lord, men worse than nothing be,
And lesse than Nothing, if compared with thee;
If lesse and worse than Nothing, tell me than,
Where is that something, thou so boasts, proud man?[22]

More often than not, the theme emerged from a consideration of the created world, humankind, or mortal life, in relation to God, the absolute being. Being mutable, fleeting and subject to the *vicissitudo rerum*, the changing nature of things, we and the natural world are nothing, as Davies made clear in 'Sic Transit Gloria Mundi':

Now This, then That, then next to This, and That,
Still changing, well I wott, t' I wot not what.
Thus is our sence deceavd, mistaking that
Which but appeares, for that which is, in deede,
And so our Sence, our Sence, doth Captivate
To mis-conceit, Corrupting Fancies Creede,
Which taks Not-beeing in true Beeings steede:
For, that is truly false what ere it is,
That is but true in show, and so is This.[23]

The shadow had a special place amongst non-entities – a paradigm of the insubstantial, the fleeting, the negative, and yet a visible phenomenon, not a figment of the imagination. As a consequence it attracted attention from the metaphysical poets. Donne's 'Lecture upon the Shadow' is one of several treatments of this topic in seventeenth-century poetry; other examples are by Henry King and, later, Thomas Traherne. Even the decidedly non-metaphysical Margaret Cavendish produced a variation on the theme, comparing shadows and echoes. But, by 1690, John Locke would be cutting through any mystification with common-sense:

> I will … appeal to everyone's own Experience, whether the shadow of a Man, though it consists of nothing but the absence of Light … does not, when a Man looks on it, cause as clear and positive an *Idea* in his mind, as a Man himself, though covered over with clear Sun-shine?[24]

In the same essay, Locke likewise dismissed the persistent controversy about vacuums, insisting that 'whatever men shall think concerning the existence of a vacuum, this is plain to me, That we have as clear an idea of space distinct from solidity as we have of solidity distinct from motion.'[25] Locke's attitude to space was that it was a 'simple idea', whose reality was therefore not in question.

The refusal of the modern empirical scientists to address ontological questions, and their determination of truth as simply the knowable or the demonstrable, left a gap which poets were filling. Questions such as whether or not, 'To be or not to be?', were bypassed by new-style scientists seeking answers to *why*? or *how*? Their project of classification and taxonomy superseded the medieval analysis of genera and species, involving categorisation and definition. Phenomenal examination would eventually dismiss the mysterious *rete mirabili*, and spontaneous generation, as fictions; it was an age in which fictions must be dismissed, an age of positivism and determination. The medieval categories and subcategories of names had established a certain hierarchical attitude to the things of the world, and indeed the celestial realm. But once the usefulness of the hierarchical concept of a 'Great Chain of Being' had been questioned, connections between at least terrestrial things were seen as horizontally ordered. The medieval rationalised view of the world was being problematised by this attention to taxonomic detail; it was beginning to be accepted that 'this' and 'that', and even 'I wot not what', could really tell us something about the world after all.

Alchemy is often seen as a proto-scientific discipline, but I would suggest that the appeal of alchemic philosophy to Donne was exactly its metaphysical aspect: its preoccupation with fundamental – or certainly elemental – questions of creation and existence. Alchemists' concerns ranged from experimental or medical empiricism to the abstractions of alchemic theory: at the core of that theory, and Kaballistic philosophy in general, was the basic opposition of metaphysics: being and nothingness. 'Nothing' or '*nihil*' was the originary, mysterious, and ultimately numinous source of created things. Hence the relevance of alchemy to Donne's poem about things which are not, 'A nocturnall upon S. Lucies Day'. Although there was a movement in English alchemy to shed its metaphysical aspects in favour of its practical applications during the first half of the seventeenth century, the alchemy of Donne's lifetime was strongly attached to belief in the existence and creative power of an immaterial substance, *spiritus*. There is no doubt that this aspect of alchemy at least became increasingly recondite in scientific enquiry during the century. As Hobbes was to report scathingly in *Leviathan*, 'common people' speak of 'aeriall substances' or spirits, but these are 'Idols of the brain', and like all idols, as St Paul said, they are nothing.[26] This sort of scientific positivism was not only anti-mystical, but effectively also anti-poetic, putatively establishing clarity of descriptions and taxonomic distinctions – a plain language of human science. Brian Vickers has shown the untenability of such a position by examining the rhetoric of the Royal Society itself, but it is evident that, conversely, spiritual and therefore often religious imagery was associated with poetic language, putatively distinguishing it from the scientific.[27]

'Reall things' and imaginary spaces

Another intriguing connection between poetic and philosophical treatment of non-entities is seen in the topic of the emotions. Davies explored the topos of love as nothing repeatedly, as in Sonnet 70:

> I cannot love no love, nor love that love
> That's like Privation, drawing neer to nought:
> That love is nothing, and can Nothing move,
> But such a something as cannot be sought ...[28]

Compare those consciously subtle distinctions between nothing and privation (the kind which drew such anti-scholastic scorn in the sixteenth century) with Donne on the same theme:

> If that be simply perfectest
> Which can by no way be exprest
> But negatives, my love is so.
> To All, which all love, I say no.[29]

Donne's use of the language of negative theology – that God's perfection could only rightly be expressed as negatives such as infinite, immense, eternal – was given the status of orthodoxy by Thomas Aquinas in the thirteenth century. This is, therefore, a metaphor taken from medieval theology, transferred from a sacred context to a profane, and like Davies' trope its point is to interrogate the reality of love. John Donne addressed the issue of being and not-being in relation to various 'idols of the brain' and non-entities: he dealt with negation perhaps most abstractly in 'Negative Love', but also spirit in 'Aire and Angels' and 'The Extasie'; soul and absence in 'A Valediction Forbidding Mourning'; grief and death in 'A Nocturnall upon S. Lucies Day'; death and hell in the Holy Sonnets.

Absence and grief were common poetic topics in the early seventeenth century – the focus of both love poems and elegies, by amongst others Robert Herrick and Richard Lovelace. The topic as dealt with by Shakespeare provides us with an interesting distinction between nothings and imaginative fictions, in Act II, scene ii of *Richard II*. There we see a remarkable dialogue on these two *négatités*, as Sartre later called such phenomenal instances of *le néant*:

> *Queen*: ...Yet again methinks
> Some unborn sorrow ripe in fortune's womb
> Is coming towards me; and my inward soul
> With nothing trembles – at which something it grieves
> More than with parting from my lord the King.
> *Bushy*: ...So your sweet majesty
> ... Find shapes of grief more than himself to wail,
> Which looked on as it is, is nought but shadows
> Of what it is not ...
> *Queen*: ... Howe'er it be,
> I cannot but be sad: so heavy sad

> As thought – on thinking on no thought I think –
> Makes me with heavy nothing faint and shrink.
> *Bushy*: 'Tis nothing but conceit, my gracious lady.
> *Queen*: 'Tis nothing less: conceit is still derived
> From some forefather grief; mine is not so,
> For nothing hath begot my something grief
> Or something hath the nothing that I grieve.[30]

These meditations on the origins, the existence and the physical effects of the emotions could be seen as precursors of the systematic exploration by René Descartes of 'les passions de l'âme' in his final work of 1649.[31] These passions of the soul were recognisably extensions of the *ens rationis* described by Suarez half a century earlier; Descartes attempted to give mechanical and (notwithstanding his reliance on a notion of animal spirits) empirical explanations of those phenomena discussed by Bushy and the Queen – trembling, fainting and weeping. Bushy's view that the Queen's fearful imaginings are 'shadows' seems to be very like Descartes' verdict on non-volitional imaginings caused solely by the body: '…their cause is not so conspicuous and determinate as that of the perceptions which the soul receives by means of the nerves, and they seem to be mere shadows and pictures of these perceptions'.[32] It is important to note that Descartes attributes a greater reality to 'imaginings and other thoughts formed by the soul' in a consciously willed manner: 'when our soul applies itself to imagine something non-existent, as in thinking about an enchanted palace or a chimera'.[33] The allusion to a chimera is significant, since the chimera was a classic scholastic example of the fiction, the nothing.[34] Suarez, a strong influence on Descartes, had stated that 'a chimera is a non-being, for if it is a fictional being, then it is a non-being'.[35] But Descartes was suggesting that we, by our wills, can make the idea real, even where its referent is non-existent.

Despite the inevitable contradictions of a *rhetoric* of empiricism, its metaphors became recurrent tropes: ones of voyaging and conquest, of colonising, mapping, penetrating and possessing. The aggressive positivism of this language of enterprise is set in relief against the contemplative, inward-looking poetics (negativism, perhaps) of much metaphysical poetry. Even inward-looking poetry could, however, appropriate the same metaphors of scientific exploration, especially through the language of anatomy, but also by treating the mind spatially. The soul was figured as a house by Davies in *Nosce Teipsum* (1599); Sir Edward Dyer announced 'My mind to me a kingdom is', to be bettered by Robert Southwell's 'My mind to me an empire is'.[36] Mere control and rule over the mind was too conservative for the seventeenth-century spirit, and even the metaphor of imperial expansionism does not properly convey the way that the journey of mental exploration itself became fetishised. One manifestation of this fetish was the mystification of the imaginative faculty, expressed recurrently in Shakespeare's work. The preoccupation of several of Shakespeare's later plays with non-entities focused on the status of the imaginary world which they constructed: the ontological status of imagined things, of fantasies, and of emotional states were all addressed, often in a metadramatic commentary upon theatrical representation itself. *The Tempest* provides a classic example: Prospero's island is on more than one level an imaginative creation, a fantasy, and within it are further magical productions – Ariel, and Caliban, who Prospero at the last confesses to be his 'child'. Ariel and Caliban are children both of Prospero's and of Shakespeare's brain. Ariel's being is ethereal, with shape but no substance, like Donne's angels, and the language used by Prospero to him – for example, 'go take this shape' – echoes the language used by Theseus in *A Midsummer Night's Dream* to describe the products of imagination:

> And as imagination bodies forth
> The forms of things unknown, the poet's pen
> Turns them into shapes, and gives to airy nothing
> A local habitation and a name.[37]

'Airy nothing' signifies the imaginary realm, the 'stuff that dreams are made on', and as Mercutio says of dreams, they are 'the children of an idle brain,/Begot of nothing but vain fantasy' (*Romeo and Juliet*, I. iv. 98–9). His shadowy realm is the nothing from which nothing (that is, fiction) comes. Prospero's island was at once an internal world, and an external projection of the imagination; at which point it becomes a metaphor for colonisation, as recent critics have observed.[38] The nothingness of imagination can be fruitful, like the original nothingness from which the world was created: nothingness is to things as imagination is to reality – they are related not in formal opposition but dynamically, as part of a movement from potentiality to actuality.

Poetic language at times negotiated both spiritualist and materialist ontologies, as in the work of Thomas Traherne or Margaret Cavendish. Metaphysical and spatial metaphors combined in Traherne's imagery of the infinity of the scope of the human mind – a negative conceptualisation of the scientific ambitions of the age. 'O what a *thing* is thought!' the poet exclaims, imaging thought as something located in an intra-mental void space: the mind is 'the empty, like to a large and vacant Room for fancy to enlarge in.'[39] The imagination is expansive, restless, a Baconian explorer of the oceans of experience: 'This busy, vast, enquiring Soul / Brooks no controul. / No limits will endure'.[40] Despite a tendency to veer towards spiritualism, Traherne's imagery owed something to the language of empirical science, and something to Cartesianism in its elevation of the capacity of thought. His metaphors of extension to describe the mind itself are in a sense anti-Cartesian, but to

Traherne the mind is an infinite space populated by unlimited thoughts, not confined by the body. There is an apparent conflation of the terms 'thing' and 'thought' in several poems – thoughts are 'brisk Divine and living Things', or 'the Things that us affect'; 'reall things', 'reall Goods', and even 'Material delights'. The notion of thought as material seems to deconstruct the opposition which had enabled Hobbes to attack 'idols of the brain' as nothings.

In spite of the differences in their intellectual perspectives, Traherne's contemporary Margaret Cavendish was similarly influenced, by both rationalism and empiricism, in her images of worlds within worlds, and that desire to concretise 'unknown realities' in mathematical terms. She played with scientific discourse constantly in her poems, and in her address 'To Natural Philosophers' made light of her own poetry when it is dealing with scientific themes, particularly the atomism which featured so largely:

> And the Reason why I write it in *Verse*, is, because I thought *Errours* might better passe there, then in *Prose*; since *Poets* write most *Fiction*, and *Fiction* is not given for *Truth*, but *Pastime*; and I feare my *Atomes* will be as small *Pastime*, as themselves: for nothing can be lesse then an *Atome*.[41]

Interestingly, the metaphor of atomism is turned on herself, recalling the self-negation of devotional poetry: 'my *Ambition* is such, as I would either be a *World*, or nothing', she declares. In a series of poems about worlds within worlds, Cavendish contemplated that there might be a whole level of being, and indeed life, too small to be seen by the naked eye.[42] The microscopic world alluded to there was another realm on the cusp between imagination and empirical fact in the seventeenth century: Leeuwenhoek's advances in the sphere were to bring him fellowship of the Royal Society in 1680. The fantasy of a miniature world in a lady's earring, developed in a subsequent poem, accentuates this theme:

> For millions of these Atomes may bee in
> The Head of one small, little, single pin.
> And if thus small, then ladies well may weare
> A world of worlds, as pendents in each Eare.[43]

Cavendish's 1666 work *The Blazing World* crosses another indistinct line into fantasy, this time utopian, where the world is on the one hand surreal and on the other a partial reflection of Restoration England. She described this herself as one of 'two worlds at the end of their poles', the other being her *Observations Upon Experimental Philosophy* to which it was a companion-piece.[44] In the preface of her fantasy work, she states that fictions are framed in the mind, a product of man's fancy, 'without regard, whether the thing he fancies, be really existent without his mind or not'.[45] In the midst of the tale, Cavendish clearly had in her satirical sights the excesses of scholastic metaphysics as well as the vain aspirations of modern science, for both seem to be conjoined in her description of the bearmen's experiments with microscopes:

> ... notwithstanding their great skill, industry and ingenuity in experimental philosophy, they could yet by no means contrive such glasses, by the help of which they could spy out a vacuum, with all its dimensions, nor immaterial substances, non-beings, and mixed beings, or such as are between something and nothing ...[46]

She there sets the vacuum up as a fiction on the same level as those metaphysical concepts of spirit, or non-being, as though in denial of the demonstrations which had been taking place during the previous two decades.

A telling and, in view of the above quotation from *The Blazing World*, ironic example of the intellectual incertitude which this essay has charted, can be seen in Margaret Cavendish's attendance of a meeting of the Royal Society in 1667. The day that she attended, the empirically-minded males, amongst whom was Samuel Pepys, were assembled to watch a demonstration of the force of a vacuum. Robert Boyle and Robert Hooke together showed that a nine-gallon, three-pint vessel weighed two ounces less after they had emptied it using Boyle's air pump.[47] When Cavendish's biographer Douglas Grant remarked that the other experiments she watched that day were more spectacular, he was perhaps missing the point: the opening demonstration would probably have been the most dramatic, particularly to the Duchess of Newcastle.[48] Cavendish was a proponent, notwithstanding her eccentric take on them, of the atomistic theories which had enjoyed a revival in the early seventeenth century.[49] Classical atomistic theory was famously antagonistic to the very idea of a vacuum in nature. What must Cavendish, with her philosophy of reliance on sense and reason for knowledge, have made of the experiment? A demonstration that one could void a space, between glass walls, of their atoms of air must have made an impact on the poet and sceptic: it is reported that she was 'delighted'.[50] Perhaps her two worlds of fiction and rational demonstration quivered momentarily at their poles, in the knowledge that speculation had become fact, the imaginary real, and that the vacuum was already an object of science, wrenched from the poetic realm of 'things which are not'.

Notes

1. Hobbes, T. ([1651] 1968), *Leviathan*, 3.34, London: Penguin, p. 428. The author is addressing the question of 'the signification of Spirit, Angel and Inspiration in the words of Holy Scripture'.
2. Boyle, R. (1660), *New Experiments Physico-Mechanical, Touching the spring of*

the Air and its effects, Oxford: H. Hall for T. Robinson. For online illustrations of his vacuum experiments, see http://www.imss.fi.it/vuoto/eesper5.html

3. Kaplan, R.L. (2000), *The Nothing That Is: A Natural History of Zero*, New York: Oxford University Press, pp. 24–5, 50–67.
4. For an account of the explanatory use of 'ether' by Hobbes, Newton and others see Rogers, G.A.J. (1965) 'Newton, the Ether, and Seventeenth-Century Science', *11th International Congress on the History of Science*, **3**, 349–54.
5. Hobbes (1968), p. 429.
6. See Hall, A.R. (1963), *From Galileo to Newton*, London: Collins, pp. 142–3.
7. See Festa, E. (ed.), *Evangelista Torricelli*, http://galileo.imss.firenze.it/multi/torricel.
8. At the Imperial Diet in Regensburg, the two metal hemispheres, having had air extracted, were attached to two teams of eight horses which, pulling in opposite directions, failed to separate the two halves of the sphere. See http://www.imss.fi.it/vuoto/eesper4.html
9. See http://www.imss.fi.it/vuoto/ for pages from the Florence Institute and Museum of History of Science on the influence of Torricelli.
10. Daunce, E. (1585), *The Prayse of Nothing*, London, sig. Giii.
11. See Erastus, T. (1923), *Disputationem de Medicina Nova*, in L. Thorndike (ed.), *History of Magic and Experimental Science*, 7 vols, New York: Columbia University Press, vol. 5, p. 688.
12. See Grant, E. (1981), *Much Ado About Nothing*, Cambridge: Cambridge University Press, pp. 116–59.
13. See Kauffeldt, A. (1965), 'Otto von Guericke on the problem of space', *11th International Congress on the History of Science*, vol. 1, **3**, pp. 364–8.
14. For a discussion of the relation between Pascal's conclusions and his, or others', experiments, see McKeon, R.M. (1965), 'Le récit d'Auzout au sujet des expériences du vide', *11th International Congress on the History of Science*, vol. 1, **3**, pp. 355–63.
15. Quoted in Colie, R. (1966), *Paradoxia Epidemica: The Renaissance Tradition of Paradox*, Princeton: Princeton University Press, pp. 252–72.
16. The analogy between early-modern and contemporary 'knowledge technology' is explored extensively in Rhodes, N. and Sawday, J. (eds) (2000), *The Renaissance Computer*, London and New York: Routledge.
17. Baudrillard, J. (1995), *Simulacra and Simulation*, trans. S. Glaser, Ann Arbor: University of Michigan Press, p. 6.
18. According to the *OED*, 'non-entity' was a term first used in the seventeenth century.
19. See Suarez, F. ([1597] 1965), *Disputationes Metaphysicae*, ed. G. Berton, 2 vols, Hildesheim: Georg Olms, vol. 2, pp. 1014–41. The plural term *entia rationis* made it more clear that individual instances, things, are being referred to, rather than abstract 'being'.
20. Davies, Sir J. (1605), *Wittes Pilgrimage*, London: John Brown, sig. C2v.
21. Davies, Sir J. (1610), *The Scourge of Folly*, London: Edward Allde, sig. A8.
22. Quarles, F. (1967), *Divine Fancies* (1632), ed. A.B. Grosart, 3 vols, New York: AMS Press, vol. 2, p. 216.
23. Davies (1605), sig. S2v.
24. Locke, J. ([1690] 1989), *An Essay Concerning Human Understanding*, ed. P.H. Nidditch, Oxford: Clarendon Press, p. 133.
25. Locke (1989), p. 179.

26. See also the parallel conclusions of Francis Bacon in his *Novum Organum* concerning the four *idola*, particularly the *idola theatri*.
27. Vickers, B. (1985), 'The Royal Society and English Prose Style: A Reassessment', in Vickers, B. and Struever, N.S. (eds) *Rhetoric and the Pursuit of Truth*, Los Angeles: University of California, p. 45.
28. Davies (1605), sig. F3v.
29. 'Negative Love', or 'The Nothing'; see Donne, J. (1985), *Complete English Poems*, ed. C.A. Patrides, London: J.M. Dent & Sons, p. 117.
30. Shakespeare, W. (1986), *Complete Works*, ed. S. Wells and G. Taylor, *Richard II*, Oxford: Clarendon Press, II, ii, 9–37.
31. See Wells, N.J. (1989), 'Objective Reality of Ideas in Descartes, Caterus and Suarez', *Journal of the History of Ideas*, **27**, 33–61.
32. Descartes, R. (1985), *Philosophical Writings*, trans. J. Cottingham, R. Stoothof and D. Murdoch, 3 vols, Cambridge: Cambridge University Press, vol. 1, pp. 325–404.
33. Descartes (1985), vol. 1, p. 336.
34. Medieval sophisms such as 'Nihil est chimæra' (Nothing is a chimera) were still being used in late sixteenth-century text books of Logic; for example, R. Goclenius (1597), *Problemata Logicorum*, Marburg: Paulus Egenolphus, Part 1, p. 220.
35. 'Chymæra est non ens; nam si est ens fictum, ergo est non ens', Suarez, ([1597] 1965), p. 1035.
36. See Davies, Sir J. (1975), *Poems*, ed. R. Krueger, Oxford: Clarendon Press, p. 10; Lucie-Smith, E. (ed.) (1965), *Penguin Book of Elizabethan Verse*, Harmondsworth: Penguin, p. 248; p. 212.
37. Shakespeare (1986), *A Midsummer Night's Dream*, V, i, 14.
38. See, for example, Brown, P. (1985), '"This thing of darkness I acknowledge mine": *The Tempest* and the discourse of colonialism', in J. Dollimore and A. Sinfield (eds), *Political Shakespeare*, Manchester: Manchester University Press, pp. 48–71; Orkin, M. (1997), 'Whose things of darkness?', in J.J. Joughin (ed.), *Shakespeare and National Culture*, Manchester: Manchester University Press, pp. 142–69; Loomba, A. (1996), 'Shakespeare and Cultural Difference', in T. Hawkes (ed.), *Alternative Shakespeares*, 2 vols, Vol. 2, London and New York: Routledge, pp. 164–91 (pp. 171–80).
39. Traherne, T. (1932), *Poetical Works*, ed. G.L. Wade, London: Dobell, p. 139.
40. Ibid., 'Insatiableness', p. 146.
41. Cavendish, M. (1653), *Poems and Fancies*, London: T.R. for J. Martin & J. Allestrye, pp. 25–9.
42. Ibid., pp. 43–5.
43. Ibid., p. 45.
44. Cavendish, M. (1994), *The Blazing World and other writings*, ed. K. Lilley, London: Penguin, p. 124.
45. Ibid., p. 123.
46. Ibid., p. 145.
47. Grant, D. (1957), *Margaret the First*, London: Rupert Hart-Davis, pp. 23–6. For a more detailed account of the visit, see Mintz, S.I. (1952), 'The Duchess of Newcastle's visit to the Royal Society', *Journal of English and Germanic Philology*, **51**, 165–76.
48. Grant (1957), p. 25.
49. For a detailed account and interesting perspective on her theories, see Rogers, J.

(1996), *The Matter of Revolution: Science, Poetry and Politics in the Age of Milton*, Ithaca and London: Cornell University Press, pp. 177–211.
50. Mintz (1952), p. 174.

CHAPTER THREE

Points Mean Prizes: How Early-Modern Mathematics Hedged its Bets Between Idealism and the World

Jess Edwards

Mathematics and textual discourse

There has been a turn, in some recent critical discourse on early-modern visual and spatial culture, against the written word. The more phenomenologically inclined of recent accounts would have us reach at least for some extra-textual, 'performative' perception in our understanding of, say, colonial geography.[1] Materialist histories of art and science seek to expose the implication of these praxes and their products in what they seem to take to be the equally extra-textual, equally performative field of commerce. Presuming her reader's complicity with a Romantic idealisation of Renaissance art, Lisa Jardine asks: 'What kind of difference does it make to our view of the Renaissance to understand that a masterpiece like Giovanni Bellini's portrait of Doge Leonardo Loredan was a strictly commercial piece of work, produced in accordance with a whole network of social obligations and power relations in fifteenth-century Venice?'[2] The difference it makes, presumably, is that we do not swallow entirely whole the frequently enunciated textual claims (made both then and now) that Renaissance art, and indeed science, was free from worldly concerns. We are growing used to being warned of the complicity of Tillyard, Burckhardt, and even post-structuralist and New Historicist readings of 'Renaissance' art with early-modern fantasies of cosmic, and by extension social, order.[3] But might we not worry that some critical materialisms threaten to bias interpretation similarly incommensurately in favour of relations of commercial exchange? Free trade, in an early-modern Europe still tangled up in pre-capitalist political knots, was no less an object of abstracted fantasy than the cartographer's or painter's abstract geometric grid. In a similar way, the modern e-capitalist dreams of the collapse of distance and the nation state, while the rest of the world lives very much at their mercy. Whilst I will not be concerned with such retrospectively conceived 'high art' as Renaissance painting, what I want to suggest here is that the original meanings of such privileged visual objects as mathematically constructed Renaissance maps, and such privileged practices as mathematical surveying, were not invested wholly in the 'high' or in the 'low'. Rather, they were determined most importantly in a vacillation *between* the ideal and the material; between liberality and use. I will argue, moreover, that the simultaneous authority and agency of those who made and owned such objects was determined in a correspondent double vision, reproduced in and through discourse and the text. I will not be looking at maps or mapping as such in this essay, but at discourses on mathematics that I believe significantly conditioned the purchase available to such mathematical applications as surveying and cartography in early-modern culture.

The first premise of mathematical double vision is Plato. The importance of Neoplatonic ideas in Renaissance visual representation of, and intervention in, the physical world has been well documented – particularly usefully, for instance, by Denis Cosgrove in *The Palladian Landscape*.[4] Cosgrove is interested in the permeation of a strain of geometric Platonic idealism, which he calls a 'Euclidean ecstasy', through sixteenth-century discourses ranging from the pure metaphysical to the determinedly practical, and through discourses concerned with newly dignified and 'intellectual' arts such as perspectival painting in between.[5] But, whilst he makes a convincing case for the investment of such practical arts as surveying with an idealist impetus, Cosgrove is not particularly interested in the mechanisms of such investment. Once again, this is a merely textual matter. In his own words, sixteenth-century writing demonstrates a 'drawing together' of the speculative and the practical; an 'alternation' between the two aspects of such a science as mathematics.[6] Writers on practical mathematics might make a 'disingenuous' claim to pure pragmatic surface, but even their practically-oriented readers were 'not unaware of the philosophical depths' beneath.[7] What I want to examine in this essay is the textual substance, even the 'literariness', of this clearly ambiguous relation of the liberal and the practical: an ambiguity which Cosgrove's spatial metaphors leave merely implicit and indeed reproduce. I want to suggest that, just as we should not elide the worldliness of such practices as surveying, and such visual objects as the map, neither should we disavow their worldiness. In fact, to do so would be (once again) to be complicit with an idealist epistemology which itself pushed narrative to the margins of truthful representation.

So to Plato: simply expressed, Plato believed the temporal world of sensibly perceived particular things to be governed by an a-temporal world of intellectually conceived universal ideas.[8] The 'real' world, as far as he was concerned, was not the natural one, but the supernatural, the former being ephemeral and illusionary; the latter permanent and true. Plato considered the objects of mathematics (triangles, circles and so on) to belong in the realm of the ideal 'real' as much as any other pure ideas, with the special qualification that, whereas part of the perfection of other immaterial ideas lay in their unitariness, instances of mathematical objects such as the triangle could be many, even infinite. In their qualified perfection, Plato allowed that contemplation of mathematical objects might serve as a preparatory study, or 'propaedeutic' for pure philosophy. Such a propaedeutic mathematics as Plato advocated is far from being the practical mathematics of the workshop, or even the procedural mathematics of the classroom and study, both of which Plato rejected utterly as contaminated with the world.[9] Indeed, Plato refused the philosophically inclined mathematician not just the physical props of a worldly mathematics (drawings of circles and triangles) but even a dynamic language that spoke metaphorically and imagined the construction of such props.[10] Such a refusal is consistent with Platonic epistemology in general, which concedes the visual and verbal figure no role in truthful thinking.

Plato's own metaphysical discourse, and subsequent philosophical discourse still heavily invested in his dualist ontology, inevitably uses props from the physical world to help the would-be philosopher think his way from the particular to the universal. Plato's tendency to use mathematical analogies to formulate his ontology may, it has been suggested, be rescued from the imputation of such discursive worldliness if we remember that his mathematics was not of the world.[11] But later Platonic discourse, beginning with the Classical Neoplatonists, took Plato's cue to expand the metaphorical investment of mathematics with metaphysics, and set the classroom mathematician's geometric point alongside other worldly analogies for existential primacy, such as light.[12] Such analogic props, according to the logic of an original Platonism, are still only that: mere props. They might, to use another Neoplatonic metaphor, be regarded as both thinking and talking 'ladders'.[13] As mere metaphors, mere images, they should be dispensed with, kicked away, on arrival at true knowledge. But in a post-Derridean world we know better than to take the supplement as 'mere' supplement. We know that the only being Plato's double world can have is through metaphor itself: through the irresolvably paradoxical, irreducibly discursive logic which says that light is both materially itself, and at the same time only a neutral, metaphorical stand-in for something better. We know that the supernatural double that flickers behind the shaky hand-drawn circle is, like the rest of Plato's supernatural realm, merely a trick of language.

Such scepticism is nothing new. Indeed, interpreting and adapting Plato's epistemology, Aristotle insisted that intellectual knowledge, including the knowledge of mathematics, had only a cognitive, rather than an existential priority over the knowledge of the senses.[14] The former, sublimated by science from experience, simply constituted more absolute and therefore better knowledge, where the latter was merely 'better known' in the sense of being more familiar.[15] But we should note that, although Aristotle's influential commentary dismantles Plato's hierarchy of being for purposes where the issue of being is explicitly addressed (in modern terms, the realm of 'philosophy'), he leaves it functionally intact where the issue is simply one of better knowledge (in modern terms, the realm of 'science').[16] He leaves the door open for his sustained valorisation of mathematical knowledge to be *misunderstood* as idealism, a possibility amply fulfilled in a whole tradition of Aristotelian commentary.[17] And many of Plato's Classical and medieval disciples, whether or not they read him through Aristotle, not only embraced his dualist ontology, but strained to make a more concrete reality out of his metaphors. Mathematical objects, they suggested, the things we 'mean' when we talk about or draw triangles, circles and so on, might be considered not just as metaphors for, or images of, the material inhabited by the ideal, but also as a third realm of being, and a stepping stone between the temporal and the absolute.[18] Sight, whose medium is light, might be considered not just as a prompt to *intellectual* memory, but also as a sense which touches upon the divine. This position was taken to an extreme by those medieval and Renaissance magi who, embracing the mystical Platonism of the Hermetic tradition, believed they could find the 'seeds' of the divine not just metaphorically but actually in nature.[19]

I am not interested in proving or even arguing for an explicit thread of conscious Neoplatonism in early-modern mathematics. Although such a thread is substantial enough, it is easily dismissed by teleological histories of science and culture as eccentric and a dead end. Even Denis Cosgrove, after making a forceful argument for the saturation of sixteenth-century Italian artistic culture with mathematical idealism, sees it giving way, 'radically undermined', in the 'seventeenth-century scientific revolution'.[20] What seems to me of more importance is the way in which Platonic *associations* persist (more or less explicitly; more or less consciously) in a whole range of early-modern texts involved in the reproduction of mathematical and visual culture. What I think we end up with, in medieval and early-modern writing influenced by a pervasive Platonic thread if not of philosophy then of discourse, is a distinct ambivalence over the status of both mathematics and the mathematical visual object: a kind of double vision.[21] Both mathematics and sight can be, within the same discourse, both transcendent and mere workaday, material analogies for the divine. And whilst there may seem in retrospect to be a great distance between the mystic 'visions' of medieval Neoplatonism and the positivist observations of the empiricist Enlightenment, it seems that, as Martin Jay has put it, the 'positive associations of geometrical order' residual in the latter still owed no small debt to the former.[22] What I would argue is that this debt is carried forward through discourse. Early-modern discourses of mathematics and visual/spatial culture inevitably combined both the materialist and Neoplatonist impulses which have been identified in rival histories as characteristic of European culture of the period. Such a combination of logically contrary impulses may not, perhaps, sit comfortably in explicit statements of artistic and scientific theory. It is, however, more likely to subsist in the more popular metaphors, parables and icons that adhere to and help to determine the meaning of early-modern geometry and vision.[23]

Where we do not acknowledge the role of such popular discourse in propagating the meaning of mathematics in early-modern culture, we are likely to conclude, as does Lesley Cormack, in one of the most influential of recent studies in this area, that mathematical interest and knowledge in the late sixteenth and early seventeenth centuries was of a 'closed and almost recondite nature'.[24] Cormack's highly useful study demonstrates the social constitution of that small body which studied mathematics in this period at the two established Universities, and joins the fairly long-standing debate about the extent and social penetration of a figurative 'Third University' in London, promoting mathematics both through closed intellectual 'coteries', and through more open attempts at popular education.[25] The implication of Cormack's work is that we should not take the mathematical knowledge evidenced in these instances as typical of early-modern society at large, and that the more ambitious claims of early mathematical popularisers to widespread influence should probably be taken with a pinch of salt. Without this pinch of salt, we are in danger of committing the materialist's Tillyardian sin of taking the elite 'world picture', perhaps beginning to be a mathematical one in the early seventeenth century, for the general one.

But where we consider the meaning of mathematics and of the mathematical visual object (be it map, picture, surveyed landscape or whatever else) to be more than a matter of knowledge narrowly and scientifically defined, then the constituency involved in reproducing mathematical meaning must surely be much broader, and the Tillyardian sin much less of a risk. Rather than just asking who knew trigonometry or how to project a map, we might also ask who knew and used the mathematical metaphors, anecdotes and stories that, in my view, gave and still give such technologies their meaning in and purchase on the world. Who has not now, or had not then, heard of Archimedes, so disinterested in the material world that he jumped naked from his bath in joy at the discovery of a formula to measure gold? Archimedes, the designer of siege engines, who was himself killed while contemplating mathematical designs in the midst of his city's siege? And who does not now, or did not then, understand such a persona and such stories to connote both the scientist's service to and freedom from the world?[26] Without any resort to explicit philosophical profundity, this story in itself transmits the powerful Neoplatonic paradox of a mathematics independent of, and yet powerful within, the world. I think that the double meaning of mathematics, and thereby its simultaneous authority and usefulness, is sustained through such popular anecdotes as this in much the same way that Plato's self-denying rhetoric sustains his illusory double world. And it is because I believe that much of the influence and effectiveness of mathematics in early-modern practical art was constituted in *discourse* on mathematics that I want to insist upon the importance of reading texts alongside such practices as early-modern surveying, and such objects as early-modern maps.

Mathematical double vision in early-modern England

In mathematical literature published in England at the end of the sixteenth century and at the beginning of the seventeenth, it is possible to identify characteristic approaches to mathematics on a continuum running between two polar positions. These extreme stances – one of a Hellenic classicism, seeking in an idealist mathematics the means for cultivating magical intellectual powers; the other empiricist, pragmatic, and closer to Latin humanism – can be identified with two giants of early-modern educational reform: the English magus and propagandist of British 'empire' John Dee, and the Huguenot martyr Pierre de la Ramée, known often to his contemporaries in the European scientific community by his Latinised name, Petrus Ramus. Both of these figures were responsible for publications which sought to broaden the audience of mathematics beyond the universities where it was felt to have fallen into neglect: Dee in his 'Mathematicall Preface' to the first English vernacular edition of Euclid's *Elements*, translated by Sir Henry Billingsley and published in 1570; Ramus in his *Geometria* (1569), and in his published lectures and addresses on mathematics.[27] Moreover, the influence of these figures was not channelled only through their own much-reprinted publications. Dee's 'Preface' attended other editions of Euclid over almost a century following its original appearance,[28] and Ramus's *Geometria*, although excluded by other popular texts from universal usage in England, was cited and sampled with great frequency for its doctrines and its organisation by other mathematical publicists.[29] A crucial point to retain, while I am setting up these figures as antagonists, is that in fact they perceived themselves as part of a unitary enterprise of reform.[30] Both intended the revival of mathematical education; both foresaw the benefits to be gained from attention to the relation between intellectual and craftsman, and published to this end. The pair corresponded as 'friends', apparently concerning mathematical texts,[31] and Ramus even sought to intervene with the Crown on Dee's behalf to gain him a university chair in mathematics which both Oxford and Cambridge at that time lacked.[32]

This spirit of cooperation between purveyors of very different philosophies of science demonstrates the social aspect of that textual alternation between practical and speculative that Cosgrove has discussed. However, textually, Dee is generally the idealist, and by and large, where he invokes the objects of mathematical study, they take their place on the supernatural side of a Platonic binary divide. Dee's 'mathematicals' are so 'absolute', and so 'free from all matter',[33] that another name than 'Geometry' 'must needs be had' for a science which 'regardeth neither clod nor turff: neither hill, nor dale'.[34] They also, however, as well as being ideal, would appear to act to join the supernatural and natural realms. '[S]urmounting all creatures', the objects of mathematics are 'used of the Almighty and incomprehensible wisdome of the Creator' as 'principall Example or pattern' so that such creatures may be 'brought from *Nothing*, to the *Formality* of their being and state'.[35] As with all Neoplatonists, Dee never gets beyond metaphor in his attempt to explain how mathematical objects can make more than an analogic connection between two sets of qualities defined through opposition. They are simply, he states, his brackets admitting implicit logical defeat, '(in a manner) middle'.[36] Yet, despite all the utterly post-Platonic optimism of Hermeticism, Dee still believes in the capacity of human art/science to imitate the divine act of mathematical bridging and generation – not just in language, but in the world. Such imitation is to be achieved through a kind of incremental shading between the material and the ideal, which bizarrely valorises not just the spiritual, but also the delicate material things which imitate the spiritual, and the delicate craft which produces them:

> Though *number*, be a thing so Immateriall, yet by degrees, by little, and little, stretching forth, and applying some likeness of it as first, to things Spiritual: and then bringing it lower, to things sensibly perceived, as of a momentary sound iterated: then to the least things that may be seen, numerable: And at length (most grossely) to a multitude of any corporal things seen, or felt: and so, of these grosse and sensible things, we are trained to learn a certain Image, or likenesse of numbers, and to use Art in them ...[37]

However much it is possible to root a modern tradition of empiricism in the ideas of Renaissance magi (and even Bacon has been traced this far) there is a perfectly distinct antimystical empirical current co-existent with the idealism of Dee and his adherents, and this is most notably represented in the figure of Petrus Ramus.[38] Where Platonism discovers the origins of mathematics in the ideal, and Hermeticism seeks to harness the virtues of these ideal origins within mundane practice, Ramus discovers a 'closed circuit' leading from and back to human use.[39] This position, which structurally anticipates the more materialist, performance-oriented modern historicisms, aligns with and extends to its logical conclusion the Latin humanist precept that theory is engendered and preceded by practice.[40] For Ramus, as for materialist histories of science, there 'is' no pure mathematics.

Ramus divides science into the successive stages of nature, art (which follows nature) and practice (which follows art).[41] In accordance with this sequence, the principles of any art are discovered not a priori, but through study of its usages and through rational reduction of these to generalities.[42] Hence, where Dee rails against the name 'geometry', Ramus famously embraces a definition of the science as 'ars bene metiendi' ('the art of measuring well') and vehemently opposes the medieval university tradition of dividing the quadrivial disciplines – geometry, arithmetic, astronomy and music – into the hierarchy of speculative and practical.[43] Ramus bases his trust both in the rightness of usage and in the possibility of knowing usage rightly on his doctrine of 'natural reason', by which he means the inherent capacity of even, and perhaps *especially* uncultivated human intelligence to follow nature, forming correspondences with the natural order of the universe.[44] This ability to follow is strictly *not* an ability to precede: there are no 'innate' ideas, as there are with Plato.[45]

The Ramist focus upon use, rather than intuited ideality, as the proper subject matter of geometry clearly feeds into what is generally viewed as the Baconian empiricist mainstream of scientific reform in the early seventeenth century. However, Ramus's commitment to induction is, like Aristotle's, limited principally to man's acquisition of science.[46] From thenceforth, once empirically proven, the Ramean sciences proceed, as with Plato and Aristotle, from general to particular. Accordingly, the geometric section of Ramus' *Scholae*, as rendered by his first English translator Thomas Hood in 1590, reproduces the essentially idealist Euclidean definitions which commence with that of the point: 'an undivisible signe in a magnitude'.[47] Behind this definition lies a qualifying gloss *not* reproduced by Hood which makes clear the distance between Ramus and Dee, and which I paraphrase here from Ramus' Latin. We *define* the point, Ramus says, as not registering physically, to our senses. We *think* of it as indivisible. However, in practice, that which is least perceptible will do.[48] Ramus' point definition, *with* its supplementary gloss, aligns with that of the highly influential English humanist Robert Recorde, who sets his agenda in *The Pathway to Knowledg* (1551) by abandoning the indivisible, 'unsensible' point of 'onelye *Theorike* Speculacion', in favour of 'that small printe of penne, pencyle, or other instrumente, which is not moved': a definition more suitable for 'practise and outwarde worke'.[49]

Pragmatic as it appears, since the material geometry invoked by Ramus and Recorde only needs, like Aristotle's cognitively primary mathematics, to be *thought* of as ideal, their disavowal of the conceptual for the purposes of practice, and replacement of it with a reduced form of the material; a materiality of least-ness, is unnecessary and, to echo Denis Cosgrove, 'disingenuous'. It keeps the conceptual, of which these writers are 'not unaware', in play: partly signified (as the conceptual can be by any material sign, however clumsy), and partly 'present', in the form of an only virtual materiality. Ramus' 'least perceptible' and Recorde's 'small printe' points enjoy the same idealist glamour (Jay's 'positive associations') as Dee's 'momentary sound iterated'. Their small-ness makes them, like Plato's mathematicals themselves, (nearly) ideal. But where Dee is obliged to trip over himself in trying to resolve his belief in mathematical perfection (only two kinds of being) and his desire for mathematical usefulness (or maybe somehow three), Ramus and Recorde's pragmatism permits them disingenuously to avoid the issue of metaphysics altogether, enjoying the glamour of geometrical idealism unabashed through simultaneously disavowing it. And those practically-oriented texts which reproduce Ramus' geometry without its qualifying glosses beyond one or two well-worn dicta are able to avoid the issue still further, maintaining a suggestive silence on the metaphysics that might ground a practical geometry. Most early-modern mathematical texts that aspire to a popular audience inevitably fall into this 'silent' category, and many seem to weave together by sheer rhetoric the potency of Dee's and the usefulness of Ramus's geometry.[50]

One example of such weaving is afforded by Thomas Hood's translation of Ramus' geometry, intended for auditors to lectures given by him under the auspices of the London merchant Thomas Smith, to an open audience composed in part of the militia got up to counter the Armada.[51] It is appropriate, given the worldly circumstances of its production, that the translation repeats Ramus' pragmatist dictum that geometry is 'the Arte of measuring well'.[52] The speech given by Hood in 1588 to inaugurate these lectures, in correspondingly humanist voice, determines the 'subject matter' of the mathematical arts as 'the world'.[53] However, parts of this speech also take on a tone very similar to Dee's then eighteen-year-old 'Preface', expressing Dee's Platonist distaste for 'clod and turff' in exhortations to his audience not to be 'looking grovelling on the ground as senceless beastes', and echoing Dee's Hermeticist desire to perceive instead the mysterious mathematical seeds of the absolute beyond/within a familiar 'outward frame'.[54] Hood's rhetoric commits, in other words, to both an idealist and a materialist-utilitarian philosophy of science.

Hood's resort to a metaphor of supplementarity – 'framing' – is typical of the promotional literature of mathematics in general, whether idealist or pragmatically inclined. In its bid to persuade readers of the value, even the necessity, of mathematics, such literature tends consistently to deploy a rhetorical manoeuvre which advocates mathematics as a necessary, even a definitive, component of human life, and yet which stresses the need for this already definitive attribute actively to be taken up. In strict Platonist terms, this taking up of mathematics should be a conscious intellectual act; in Ramist terms, it might often be an unconscious practical one. Ramean man enacts his mathematical nature not by contemplating the perfection of the geometric figure, but simply by going about his everyday mercantile affairs, and even by using a pair of scissors. And yet mathematical promoters performing their sales-pitch rhetoric of mathematical supplementarity can seem once more to hedge their bets. At one point in the speech inaugural to his mathematical lectures, Hood tells the well-known mathematical fable of the Greek philosopher Aristippus who, when shipwrecked on the shore of Rhodes, was relieved to find the 'signs of man' in geometric figures sketched on the sand: traces which Hood's retelling of the tale calls 'footsteppes'.[55] Although these figures must have been made *knowingly*, with an intellectual *knowledge* of mathematics – and Aristippus was, after all, a philosopher – Hood's metaphor also implies that man 'practices' mathematics as naturally; as un-self-consciously; as habitually, we might say, as he leaves a footprint.

Petrus Ramus himself considered the human condition 'miserable' without mathematics, not merely embellished with it, precisely because of the 'everyday' aid, manifest in the most prosaic tasks and instruments, with which it provides us.[56] But this essentially democratic vision of a world of mathematical practice, perpetually observed and informed through an educative loop which never took its mind off the end of practical 'use', did not survive long beyond the utopianism of the earlier seventeenth century. And the extreme alternatives of Dee's idealism and Ramus' pragmatism, both of which promoted mathematics as necessary to human life, were survived by a kind of 'third way' promotional rhetoric which disavowed philosophies of mathematics altogether, and resolved the paradox of the necessary human adjunct into another metaphor of supplementarity: the metaphor of ornament.

The origins of a notion of ornamental, rather than necessary, mathematical use can be traced to the original humanist reforms in English education set in motion by Erasmus, Vives and Ascham, which promoted literary disciplines modelled on revived Latin texts. This Latin humanism, in contrast to the speculative Hellenism of the Florentine Renaissance, showed antipathy towards any but a 'moderate' immersion in the sciences, and then only those geared to 'som good use of life', as Ascham put it.[57] Such 'polite learning' (in Mordechai Feingold's words) as the English humanists did countenance was required to fall short of the changing of 'mens maners' that was thought by Ascham to attend too much contemplative, theoretical mathematics.[58] In such a climate those who supported scientific education were obliged to disavow its profundity; its tendency to foster withdrawal from 'the world'. Sidney felt moved to promise his tutor that he would only 'peep through the bars' of geometry.[59] On the other hand, however, most university scholars were not prepared, as Ramus avowedly was, to sacrifice the liberality of a mathematics divested of its unworldliness. Such out and out pragmatism was resisted not just by mystics and unworldly scholastics such as Ramus' sworn enemy the Jesuit Charpentier, who (ironically, given his name) called applications 'les ordures et fientes des mathématiques' ('the gross and excremental aspect of mathematics'),[60] but also by literary humanists who found their ancient sources snobbish about the purely practical.[61]

Despite the undoubtedly practical, or utilitarian, turn of later mathematical material, Charpentier's recoil from application as the waste, or excrement of mathematics, persists. It is present in such disavowals as Sidney's as a muffled counterpoint to mathematical pragmatism, holding mathematics and the mathematician back from the ignobility of utilitarian necessity, without risking the opprobrium cast upon utter disengagement. Where distinct idealist and pragmatist attitudes are combined through areas of structural resemblance in other (principally earlier) writing, in later commitments to 'polite learning' they are fused unacknowledged into a single amalgam, whose very weightlessness elides their metaphysical incompatibility. This weightlessness is neatly characterised in *Mathematical Recreations*, a French treatise translated into English by the notoriously anti-utilitarian mathematician William Oughtred in 1633. Rather than exclusively at the head of immutable Euclidean commandments, this treatise locates the first principle of geometry amongst a cache of trifles. '[S]undry fine wits', observes the author, 'as well amongst the Ancient as Moderne, have sported and delighted themselves upon severall things of small consequence, as upon the foote of a *fly*, upon a *straw*, upon a *point*, nay upon nothing.'[62]

Essentially, I think, polite mathematics newly encodes the vacillation between transcendence and engagement that we can see effected in various other ways through a long tradition of mathematical discourse. Its precarious balancing act between Ramus' unknowing use and the unuseful knowledge of classical Platonism, its affected nonchalance regarding both contemplative science and profitable craft, hides a silent investment in both. Sometimes such investment shows through the clever rhetoric required to maintain this balance. Conceding that his worldly audience may be impatient to see in mathematics 'some use, or some commoditie fit for a common weale', Hood parries by undermining the utilitarian notion of profit and returning to the lighter ornamental.[63] '[W]hat profit hath the Diamond on your finger, for which you gave an hundred pound?' he teases: 'You can not answer me at a blush, except you aleadge your pleasure alone.'[64] 'I would not have you measure eche thing by the price', claims Hood, playing to his audience's vanity as adeptly as Shakespeare's Mark Antony, 'for commonlye schollers are not covetous men.'[65] He bluffs here rather cynically, of course, since diamonds are appreciated by merchants precisely for their price. The scholar's renowned uncovetousness being a function of his contemplative unworldliness, the cynicism of Hood's rhetoric is made still more apparent when, a little later, he once more avoids laborious talk of illiberal use through addressing his audience in their familiar professional, rather than their projected liberal, guise. 'If you your selves were not Merchant men', Hood flirts again, 'I would tel you what proffit you reape heerby, but your dayly experience saveth me that labour.'[66]

Hood does not *talk* about profit here, just as Mark Antony does not praise Caesar: he does not have to. Instead he speaks of diamonds, both a source of purely visual pleasure *and* a highly profitable commodity, and leaves gaps for the merchants' commercial pipe-dreams to fill. He also, moreover, allows the merchants to enjoy fantasies of themselves as liberal scholars, and not just scholars of polite diversions. For besides the profit that he need not mention there is something else that lies behind the pleasant facade of polite mathematics: 'some notable thing ... that is not common to other artes, els could it not breed such exceeding joy in the lovers thereof, as we see it dooth'.[67] And so we return to disinterested pleasure: to the Platonic consolation that secretly grounds the facile liberality of polite mathematics; to the philosophical depths that early-modern mathematical publicists disingenuously affect to disregard. Hood's account of the 'joys' of mathematics, instanced with reference to two famously, even definitively, disinterested mathematicians, denounced as such by Ramus, is immoderate, rather like Dee's more purple passages. It is even ridiculous; and is interesting precisely for its *over*emphasis of the useless, in compensation for a seemingly utilitarian agenda. In this company, the solid pragmatic reputation of Hood's first instance Thales (a merchant and a practitioner, who reputedly founded geometry by seeking the principles of usage) cannot hold down the balloon. '*Thales*', Hood recalls, 'offred up an Oxe for joy, and *Plato* an hundred having found out certaine Geometricall conclusions. *Archimedes* sifting out the Goldesmithe's deceipte in making of King *Hiero* his Crown, could not contain him self for joy, but ran as he was stark naked through the street.'[68]

Faced by an audience of merchants, speaking from within the third, and eminently most worldly, 'university' of London, and promulgating the mathematical teachings of that prophet of Protestant utilitarianism Ramus, Hood still cannot let go of the absolute. It is obvious why not. He wants (like Archimedes in the fables) both the authority of liberal transcendence and the agency of worldly use. Why not reverse Jardine's question cited at the beginning of this essay? What if the most ordinary, pragmatic and commercial aspects of early-modern practical art were invested with a spirit of mathematical idealism? And what, to answer the anti-textualists, if we couldn't appreciate this investment (any more than the commercial one) except by reading texts? I have not at any point in this essay dealt with the kind of mathematics being done by seventeenth-century mathematicians, whether at the level of 'pure', or 'procedural' mathematics, or at the level of application. Rather, I have examined the textual evidence of early-modern discursive attempts to give mathematics and its applications meaning. I have suggested that if mathematics transmits 'authority' it does not do so just, or even principally, in the form of techniques, formulae or even representations (maps, pictures), but through the twists and turns of discourse such as Thomas Hood's, which tries and fortuitously fails to say what mathematics is.

Notes

1. See, for instance, Seed, P. (1995), *Ceremonies of Possession: Europe's Conquest of the New World 1492–1640*, Cambridge: Cambridge University Press.
2. Jardine, L. (1996), *Worldly Goods*, London: Macmillan, p. 28.
3. This is the predominant editorial tone of, for instance, Cosgrove, D. (1999), *Mappings*, London: Reaktion.
4. Cosgrove, D. (1992), *The Palladian Landscape*, Leicester: Leicester University Press.
5. Ibid., ch. 8.
6. Ibid., p. 211; p. 209.
7. Ibid., p. 213; p. 209.
8. For an account of Plato's two divided realms, see Merlan, P. (1960), *From Platonism to Neoplatonism*, The Hague: Nijhoff.
9. For an account of the differences between what he regards as 'proper' mathematics and what he calls Plato's 'meta-mathematics', see E.W. Strong (1966), *Procedures and Metaphysics: A Study in the Philosophy of Mathematical-Metaphysical Science in the Sixteenth and Seventeenth Centuries*, Hildesheim: Georg Olms.
10. See Wedberg, A. (1955), *Plato's Philosophy of Mathematics*, Stockholm: Almquist and Wiksell, p. 108.
11. See Strong (1966), pp. 17–19 and 26–7.
12. See, for instance, Proclus (1970), *A Commentary on the First Book of Euclid's Elements*, trans. and ed. G.R. Morrow, Princeton, NJ: Princeton University Press, p. 133.
13. Denis Cosgrove (1992, p. 210) notes use of this metaphor, along with that of the 'key', in popularising, but still Platonically inspired, early-modern mathematical literature to signify the provision of access for 'practical' men to previously esoteric and patrician mathematical discourses.
14. See Strong (1966, p. 11) for an account of what he calls Aristotle's 'moderate realism', which – by way of contrast with later 'excessive realisms' – believed only in the greater clarity of mathematical knowledge, rather than its participation in the absolute intellectual truth of Plato's metaphysical 'real'.
15. See Heath, T. (1949), *Mathematics in Aristotle*, Oxford: Clarendon Press, p. 5, for an account of absolute and familiar 'better knowledge' in Aristotle.
16. See Merlan (1960), p. 75.

17. Ibid., pp. 62–83.
18. See Merlan (1960), for a full account of 'excessive realism', or – as he calls it – 'realistic trialism', from the early Christian centuries to the Renaissance.
19. For an account of Hermeticism, see French, P. (1972), *John Dee: The World of an Elizabethan Magus*, London: Routledge.
20. Cosgrove (1992), p. 192.
21. Jay, M. (1994), *Downcast Eyes*, Berkeley, CA: University of California Press. Jay's archaeology of the foundations of 'iconophobia' in twentieth-century French thought, provides several instances of early-modern double vision in its first chapter, titled 'The Noblest of the Senses'.
22. Ibid., p. 54.
23. See ibid., pp. 52–64 for a rehearsal of the rival claims of Florentine perspectivism (idealist) and Dutch 'descriptive'art to have helped constitute the denarrativised modern scientific gaze.
24. Cormack, L. (1997), *Charting an Empire: Geography at the English Universities, 1580 –1620*, Chicago: University of Chicago Press, p. 124.
25. Ibid., p. 103.
26. Ibid., p. 95n, for reference to the generally Platonist uses and interpretations of Archimedes' nakedness, but also one instance where it signifies his unadorned practicality.
27. See Hooykaas, R. (1958), *Science, Humanisme, et Réforme: Pierre de la Ramée (1515–1572),* Leyden: E.J. Brill, for a thorough account of Ramus' mathematical works. See also Nuti, L. (1988), 'The Mapped Views by Georg Hoefnagel: The Merchant's Eye, the Humanist's Eye', *Word and Image*, **4** (2), 545–70.
28. French (1972), pp. 172–7.
29. Hooykaas (1958), pp. 106–18.
30. French (1972), p. 167.
31. Ibid., p. 142.
32. Hooykaas (1958), p. 105.
33. Dee, J. (1650), 'Mathematicall Preface', in Euclid, *Euclides Elementes of Geometry: The First VI Books*, trans. T. Rudd, London, sig. B2v–B3v. This later appearance of Dee's 'Preface' differs only typographically from the original.
34. Ibid., sig. E2r.
35. Ibid., sig. B3v.
36. Ibid., sig. B2v.
37. Ibid., sig. B4v.
38. See Feingold, M. (1984), *The Mathematician's Apprenticeship: Science, Universities and Society in England, 1560–1640*, Cambridge: Cambridge University Press, p. 4, for a general summary of the occultist thesis in the history of science, as championed by Frances Yates.
39. Hooykaas (1958), pp. 20–21. Any translations from Hooykaas are my own.
40. Ibid., p. 21.
41. Ibid., p. 22.
42. Ibid., pp. 23, 51.
43. de la Ramée, P. (1569), *Scholae Mathematicae*, London, p. 1; Hooykaas (1958), p. 25.
44. Ibid., p. 51.
45. Ibid.
46. Ibid., pp. 52–4.
47. de la Ramée, P. (1590), *The Elementes of Geometrie. Written in Latin by That Excellent Scholler P. Ramus*, trans. T. Hood, London: n.p. [p. 1].

48. de la Ramée (1569), p. 5.
49. Recorde, R. (1551), *The Pathway to Knowledg*, London, sig. A1r.
50. See Feingold (1984, pp. 13–21) for an account which characterises the early seventeenth-century scientific community as 'protean and heterogenous' rather than divided into rigorously aligned camps.
51. Hooykaas (1958), p. 117.
52. de la Ramée (1590), p. 1.
53. Hood, T. (1974), *A Copie of the Speache Made by The Mathematicall Lecturer ... at the House of Thomas Smith*, Amsterdam: Theatrum Orbis Terrarum, p. 22.
54. Ibid., sig. A3v.
55. Ibid., sig. B3v. For the version of this tale which most ensured its early-modern penetration see Vitruvius, M.P. (1960), *The Ten Books on Architecture*, trans. M.H. Morgan, New York: Dover. Vitruvius' own telling of the tale suits the Platonist environment of Renaissance Rome into which his 'rediscovered' text was introduced with its 1486 publication.
56. Hooykaas (1958), p. 82.
57. Quoted in Feingold (1984), p. 30.
58. Feingold (1984), p. 29.
59. Quoted in Feingold (1984), p. 191.
60. Quoted in Hooykaas (1958), p. 78.
61. Hooykaas (1958), p. 61.
62. Leurechon, J. (1633), *Mathematical Recreations*, trans. W. Oughtred, London, sig. A2r.
63. Hood (1974), sig. B1v.
64. Idem.
65. Ibid., sig. B2v.
66. Ibid., sig. B3v.
67. Idem.
68. Idem.

CHAPTER FOUR

Bantering with Scripture: Dr Archibald Pitcairne and Articulate Irreligion in Late Seventeenth-century Edinburgh

David E. Shuttleton

The social history of the seventeenth-century 'Scientific Revolution' tends to focus on the activities of the Royal Society and Stuart patronage of virtuosi culture in England immediately after the Restoration. Less attention has been paid to Stuart patronage of scientific activity beyond London, and the subsequent impact of the Revolution Settlement of 1688–89 with regard to changes in patterns of private and institutional patronage. This chapter considers virtuosi medical culture in Edinburgh, in particular after the distinctive terms of the 1689 Settlement in Scotland established the Presbyterian Kirk and inaugurated an associated purge of Episcopalians from the Scottish universities. This reopening of old sectarian wounds took place within an increasingly more organised, professional and academic medical context in which personal and political rivalries generated rancorous debates over medical theory and practice. We are reminded of Roy Porter's observation, as quoted by Claire Jowitt and Diane Watt in their introduction to the present volume, that attention to seventeenth-century scientific culture reveals the co-existence of 'all manner of rival knowledge claims, and multiple models of epistemological legitimacy', behind which lie 'rival ideologies' promoted by disparate religious, national, and factional interests. In the same place my editors note Charles Webster's observation that 'any understanding of the seventeenth-century world view necessitates attention to the character of religious motivations ... and of involvement in contemporary political affairs'.[1] In the light of such approaches, those historians of science who have already begun to explore the ideological nature of the factionalism between the medical virtuosi of late seventeenth-century Edinburgh have drawn attention to the social context of such rivalries as illustrative of how the highly rhetorical claims amongst natural philosophers for epistemological authority were deeply interested. As Anita Guerrini remarks, we find that 'intellectual questions are intimately bound up with questions of status and behaviour, and these problems are revealed in a debate over language': Are physicians scholars or artisans? Should they have medical degrees and be licensed? Should they write in Latin or English?[2] As we shall see, in this contest for scientific authority the charge of being irreligious served all sides as a convenient means of demonising one's professional opponents.

This chapter focuses on the charges of irreligiosity, blasphemy and atheism levelled against the Edinburgh physician, mathematician, Newtonian medical theorist and poet, Archibald Pitcairne (1652–1713); charges that are in themselves symptomatic of the heated and bitter sectarian divisions that continued to shape Scottish society after the Interregnum. By the time of his death Pitcairne was considered by many to be one of Britain's leading medical theorists.[3] During the 1690s he played a central role in the so-called 'Edinburgh Fever Dispute'.[4] As reflected in more than 30, often acerbic pamphlets, this controversy even prompted a so-called riot between rival factions in the recently established Royal College of Physicians, Edinburgh (RCPE). As a well-documented case-history in scientific controversy, this local medical dispute illustrates how differing claims for explanatory knowledge are expressions of rival power interests, for, as Guerrini observes, 'in terms of theory, each side in the debate claimed expertise in the new mechanical philosophy' but the 'real issue was therapy: not so much in specifics, which varied only slightly, but how it was derived'.[5] Whilst Pitcairne made claims for a mathematical, ostensibly 'Newtonian' and above all learned, theoretical basis for his practice, his opponents defended the empirical methods of Thomas Sydenham.

The Edinburgh dispute did not occur in isolation. The leading Scottish medical virtuosi corresponded with fellow physicians and natural philosophers at London – where some of their pamphlets were surreptitiously printed – and with scholars at other centres of scientific enquiry throughout Europe. Whilst Pitcairne and his key rival in these Edinburgh controversies, Sir Robert Sibbald (1652–1724), were both members of the Royal Society and had been co-founding members of the RCPE in 1681, nevertheless, their sometimes mutual efforts to organise related virtuosi projects and institutions in Edinburgh, though encouraged by James VII when he visited the city in 1681, were soon to be destabilised by the efforts of Williamite agents sent to purge Episcopalians from the Scottish universities after the 1689 Revolution reestablished the Presbyterian Kirk. As an early champion of the application of corpuscular and iatro-mathematical theories in medicine, Pitcairne's attempts to introduce Newtonianism into the Scottish university curriculum left him branded a blasphemer and atheist. As emerges below, unlike Newton – a Whig, who in the 1680s publicly opposed the Catholic James VIIth's attempts to have placemen at Cambridge, and who was to die a pillar of the Hanoverian establishment – Pitcairne's later career reveals how claims for Newtonian methodological and epistemological innovation could remain rooted in Jacobite political conservatism. Nevertheless, Pitcairne participated in the emergence of a distinctively modern cultural topography in which certain key, predominantly male homosocial sites of intellectual exchange become the focus for Kirk vigilance. Most readily policed were the university regents, their syllabus and library. Beyond the academy similar factionalism characterised the fledgling RCPE, as virtuosi debate spread to such informal spaces as coffee houses, semi-private theatres and book auctions. Such heterodox sociability accords with the ideas of Jürgen Habermas, who influentially associated the high Enlightenment of the following century with just such increasing intellectual debate within the bourgeois public sphere.[6] Reputedly running his practice from an Edinburgh howff (a public wine-shop), for later generations Pitcairne epitomised the physician as Restoration wit. But, whilst rival physicians might still sometimes cross swords, by the time of his death near the close of the reign of the last Stuart, Queen Anne, Pitcairne's notoriously caustic, witty manner was already being repudiated in favour of the more polite codes of civility associated with this longer-term cultural process; codes encouraging a more moderate ecumenicism summarised in the famous call of Addison in *The Spectator* to bring 'Philosophy out of Closets and Libraries, Schools and Colleges, to dwell in Clubs and Assemblies, at Tea-Tables and in Coffee-Houses'.[7]

Recent years have seen a reassessment of Scottish Enlightenment culture, pushing its emergence back into the late seventeenth century. As Douglas Duncan summarises, 'the late seventeenth and early eighteenth centuries, previously shunned as barren wastes, have become a focus of specialised scholarly investigation, mostly aimed at autonomous areas of the nation's life – education, the established church, and the study of the law – which immediately affected Enlightenment thinkers'.[8] With Roger Emerson's reconstructions of this virtuoso activity, we should add antiquarianism, natural philosophy and, in particular, medicine to this list.[9] Both Pitcairne and Sibbald entertained religious doubts, but in a plethora of contestatory literature ranging from essays and pamphlets to orations and satirical poems Pitcairne is a particular target for the charge of irreligion. Emerson comments upon his significance as a thinker 'whose God was certainly made necessary by the requirements of Newton's physics more than by the traditional arguments for the being and attributes of God', and for whom, as a 'natural religionist Christianity may not have been essential'.[10] John Macqueen asserts that through the pursuit of natural philosophy 'Pitcairne became virtually a free-thinker'.[11] Duncan goes even further declaring that 'perhaps ... Pitcairne's radical scepticism helped to make Hume possible in Scotland, or at least to facilitate the metaphysical enquiries of young Edinburgh intellectuals in the Rankenian Club soon after his death'. However, he also notes the 'strongly conservative and reactionary' elements in Pitcairne's intellectual and political orientations, which render categorising his place within any such genealogy difficult.[12] Similarly Emerson observes that, although Pitcairne 'was hardly an ordinary or orthodox Episcopalian', when faced with conflicting accounts of him as deist, atheist and Socinian, his 'beliefs are difficult to unravel'.[13] Whilst clarifying these claims, this essay is ultimately less anxious to establish a core set of authentic beliefs for Pitcairne than to trace the cultural and political contexts within which the charges of irreligion operated.

Scholar and blasphemer

In Pitcairne's career, theoretical innovation was in part a product of his political conservatism. As an Episcopalian-Jacobite, in September 1685 he had been nominally appointed one of the first professors of medicine in the newly formed Faculty of Medicine in the Town College (later Edinburgh University) by James, Duke of York, as part of a deliberate attempt on the part of the future James VII to promote a pro-Royalist intelligentsia in Scotland.[14] It is, however, unlikely that Pitcairne delivered any public lectures in this capacity since he was politically marginalised by the Revolution, after which the Presbyterian Council proceeded to enforce the Act of Visitation requirements for university regents to swear oaths of allegiance to William and Mary. After 1689 Pitcairne openly maintained his Episcopalian and Stuart loyalties that precluded taking any such oaths. With some legal training, he soon found himself having to defend his friend Professor David Gregory and other regents against charges of profane behaviour and heretical teachings. A moral clean-up provided a convenient excuse for the newly established Kirk to remove political enemies as well as demonise potentially threatening innovations in the teaching of natural philosophy.[15]

In 1691 Gregory fled to Oxford whilst Pitcairne briefly held a visiting chair of medicine at Leyden (1691–92).[16] As Guerrini has shown, Pitcairne's Leyden lectures – eventually printed as *Dissertationes Medicae* (1701) – were internationally influential in launching the purportedly Newtonian iatro-mathematical project which Pitcairne continued to pursue after his return to Edinburgh where he allied himself with the Incorporation of Surgeons in moves to establish an anatomy theatre.[17] He also satirised Presbyterian fanaticism in the Hudibrastic poem *Babell* (1692), and in a closet-drama *The Assembly* (written 1692?, published 1720).

Accounts of Pitcairne's significance as a reputed sceptic have been tentative for, as Nicholas Phillipson observes, the matter of religious scepticism in Scotland 'is a subject waiting to be opened up'.[18] In Pitcairne's case this is hampered by the lack of any full, modern account of his varied roles as medical theorist, Latinist poet, playwright, satirist, pamphleteer, bibliophile, antiquarian and academic patron. In this chapter I can only address the accusations of irreligion and atheism levelled at Pitcairne, a project which gains impetus from an alarming, historically-contingent episode: the execution of the 20-year-old Edinburgh student Thomas Aikenhead for blasphemy on 8 January 1697. Drawing together the trial evidence, Michael Hunter has reconstructed and contextualised this notorious case, which is often cited as illustrative of the extreme state-sanctioned religious bigotry active in Williamite Scotland.[19] He finds it symptomatic of 'the marked anxiety about heterodoxy to be found in Scotland in the 1690s, in the context of a sustained attempt by the Kirk to achieve a truly godly society'.[20] Pitcairne gets a passing, though key, mention when Hunter alludes to 'the likely cause of such disquiet' being 'provided by the attitudes and activities of the doctor and intellectual Archibald Pitcairne and his circle'.[21] Any direct biographical connection remains elusive, but the fact that Aikenhead's father was a prominent Edinburgh apothecary, and that the ill-fated son was a student in the university certainly suggests the physician's possible influence.[22] We are thus prompted to consider more closely Pitcairne's role within this heightened climate of state-sanctioned moral cleansing.

Pitcairne's contemporaries were sharply divided over his religiosity but the fact that he generated highly conflicting statements is indicative in itself of his centrality for understanding the ideological tensions shaping late seventeenth-century Scottish virtuosi culture. At the time of his death, aged 61, on 20 October 1713, many confirmed that Pitcairne had been an eminent and charitable physician, but few agreed over his Christian adherences. Writing to a fellow Presbyterian minister three days before Pitcairne died, the puritan scholar-minister Robert Wodrow (1679–1734) noted:

> I'm very apprehensive of some sad judgement on these lands, atheism and irreligion are so prevalent, notwithstanding of the Gospell light we have been privileged with. That miserable creature D. P[itcair]n has of late been very bussy for the D[ivel]'s interest among the people, but is now confined to his room and it's thought cant last long.[23]

The covenanting Wodrow represents an extreme, but since Pitcairne had become the demonised focus for such fears of a national decline in faith his adherents were equally anxious to emphasise that their patron had died at one with God. Dr John Drummond, for example, wrote to inform Pitcairne's protégé Thomas Bower, Professor of Mathematics at King's College,

Aberdeen, of their mutual patron's death 'to the great detriment of learning, and the vast regret … of all learned men, both at home and abroad':[24]

> all the while he kept house, he was in the greatest tranquillity and composedness of mind imaginable; and after this manner, without pain or trouble, and with just apprehensions of God and religion, as he constantly lived, left the world; … The calumny of Atheism objected against him by fanatical and enthusiastical spirits, who brand both good and learned men with this odious name, that cannot come up to their ridiculous opinions and notions, is absurd and false; for no man believed more firmly in the existence of a God, and demonstrated it more clearly … [He] has proved against Des Cartes and others, that an animal cannot be mechanically produced, but must owe its original to a Supreme Being.[25]

Drummond thought Pitcairne 'one of the greatest geniuses that this age has produced', but when Wodrow records Pitcairne's death, though he reluctantly acknowledges him as 'the most celebrated physitian [*sic*] in Scotland', he paints a very different picture:

> some people talk y't every day he did read a portion of the scripture, tho' it seems he made ill use of it. He was a professed deiste [*sic*], and by many alledged to be ane atheist, tho' he has frequently professed his belife [*sic*] of a God, and said he could not deny a providence. However he was a great mocker of religion, and ridiculer of it: he kept noe publick society of worship: on the sabbath had his sett meetings, for ridiculing of the scriptures and sermons. He was a good humanist, and very curiouse in his choice of books and library. He gote a vast income, but spent it upon drinking, and was twice drunk every day … Some say he had remorse at his death; but others that he continued to mock at religion, and all that is seriouse.[26]

These conflicting, typically voyeuristic, deathbed representations should be read in the particular light of Bishop Gilbert Burnet's popular 1680 report on the deathbed conversion of another sceptical poet-wit, John Wilmot (often reprinted as *A Mirror for Atheists*).[27]

Charging Pitcairne with atheism was therefore largely a rhetorical gesture amongst his ideologically divided contemporaries. Interested anecdotes are just that, but Wodrow's prejudicial account nonetheless suggests the key elements in the Pitcairne mythos: his Jacobitism, competitive professionalism, and leadership of a dissolute 'club', a reputation for mocking Kirk elders if not scripture, and his controversial iatro-mathematical theories and factionalist activities as pamphleteer, poet and satirist. We are faced with at least two versions of Pitcairne: either a drunken promoter of atheism who publicly mocks religion and corrupts others (including Aikenhead?), or a scholar who had arrived at a considered philosophical position as a rational deist. Pitcairne's medical works reveal him as a follower of Descartes and the Italian iatro-mechanists, but he never endorses outright materialism, nor argues for atheism. Indeed, one anecdote, to which Wodrow alludes, 'of ane apparition he [Pitcairne] had frequently, qch he ouned' records Pitcairne's reputed belief that a pact he made with one David Lindsay, that if either died he should visit the other as a spirit, was fulfilled when the latter's spirit took to appearing to Pitcairne at Edinburgh after his friend's death in Paris.[28] Although this could be read as a tale put about by the more extremist of Pitcairne's puritan opponents eager to bring charges of witchcraft, it might equally be seen as an example of the untroubled co-existence, noted by my editors in their introduction, of what we have since falsely wished to polarise as oppositional scientific and superstitious explanations amongst seventeenth-century natural philosophers.

'… a Number of extravagant, insolent, mathematical Atheists'?

Pitcairne's enemies frequently denounced him as the heathen figurehead to a notorious 'club' of physicians, advocates and students. They met in the smoky basement Edinburgh howff (wine-shop), known affectionately as 'the Greppa' or 'Grepping Office', where Pitcairne reputedly based his practice and which he celebrated in a poem which deifies the landlady, one Mistress Henderson, as Prosperine.[29] In contrast, Pitcairne's enemies portrayed this 'cell beneath the ground' as being as 'Dark as his writings, as his schemes profound'; a Hades where 'discord, rage and malice dwell' and where the drunken physician exerts a devilish, atheistical sway over his obsequious 'underlings':

> Thus mett, they satt them down each in his place
> And each his eyes cast on their masters place
> Eager to know, what twas he would like best,
> Whether he'd have them lye, blaspheme or jest
> Or scold mankind, or mock the deities
> Or plot some mischiefe, or admire his lies.
> All ware prepar'd to Signallise their Zeal, –
> By som [*sic*] vile slander, lately forg'd in Hell
> And to applaud, att every wanton Tale.
> O! how they'd jeir the Doctors of the Town …[30]

As his own letters reveal, Pitcairne's reputation for drunken conviviality was well earned, but in this respect he merely conformed to a dominant mid-century cavalier mode of manliness.[31] Pitcairne's club was clearly a relatively open social forum for the discussion of natural philosophy, literature and historical criticism. The 'Grepping Office' may have provided the venue for the 'meetings' specifically 'sett up' to mock religion, to which Wodrow refers, but his charge that they met on Sundays hints at some inner-circle meeting at Pitcairne's house. This club was certainly a direct precursor of the poet Allan Ramsay's 'Easy Club', and other polite gentlemen's societies of the Edinburgh Enlightenment.[32] Given Pitcairne's international scholarly connections, his circle represented one of the most cosmopolitan intellectual groupings in late seventeenth-century Scotland.

At a time when one of his patients, a young girl, was being fed a magpie pie by her family to ascertain if she was bewitched, Pitcairne was seeking to place medicine on as firm a mathematical footing as Newton was doing with astronomy, chemistry and physics. When Pitcairne dissected a 'negro' soldier who 'dy'd of the small pox' in 1694 he observed that 'his gall was copious & absolutely black' and wondered if his 'colour might be icterus niger? and this the natural reward of the Canaanits [*sic*] original Sin'.[33] But whilst anatomical observations might ultimately raise theological questions in fact, as Guerrini argues, Pitcairne's purportedly Newtonian project was rather lax with regard to verifiable empirical evidence. The promotion of iatro-mathematical explanation stemmed from a desire to maintain traditional learned authority against potential erosion by upstart empirics and quacks, rather than from any pure spirit of anatomical discovery. The apparent sureties of elitist mathematics served to uphold the natural and social order against the Whig-Presbyterian threat to established intellectual and political authority.[34]

Presbyterian hostility to mathematical explanation is targeted in those sections of Pitcairne's satirical comedy *The Assembly* specifically aimed at the 'Uproars and Tumults' provoked by the 'arbitrary' government of the Town College under its new principal, the staunch Presbyterian minister, Sir Gilbert Rule. In an exchange between 'Mr Novel' (a 'Jacobin' newsmonger) and his 'fanatick' counterpart, 'Abednego Visioner', the former observes that now many students have 'more Mathematicks and Philosophy than their Regents, who know nothing but metaphysical jargon'. When Abednego suggests that ''tis better to go to a private Chamber, than to be abus'd by a Number of extravagent, insolent, mathematical Atheists', we detect the provocative voice of Pitcairne himself in Mr Novel's response that, on the contrary, 'all are Atheists except Mathematicians'. Abednego is incensed:

> O Intolerable Impudence! Shew me a mathematician among a Hundred that cares for the Confession of Faith? I'm told that the first Proposition of Euclid is to prove that the World is eternal; and the second, that there is not a God; Besides, one must have a Compact with the devil 'ere he can understand them. I put it to Trial, and upon my Honesty I cou'd not learn to speak one word of them.[35]

This self-righteous antagonism echoes that displayed earlier in the play by Rule himself, as satirised in the figure of 'Salathiel Little Sense', whose bombastic speeches are characterised by bad Latin and cod-scholastic terms like '*Categorimatice* and *Symategorematice*'. With satisfaction Salathiel tells The Assembly that the now purged College is sending forth students who have been well-taught to 'despise vain Philosophy and Mathematicks'. They have been instructed 'in many things which the Malignants, who want grace, say are Contrair to reason', but which are in fact above 'carnal Reason'.[36] Mr Novel observes mockingly that Salathiel is 'now upon a project of making a *German Randy Beggar*, extraordinary Professor of theology'. Salathiel himself has already expressed satisfaction that; 'It's the Lord's Doings, that hath purged the Fountains and Seminaries … overgrown with [Des] Carte's Mathematicks and humane Reasoning; yea some of them were so blasphemous, as to maintain that the King was supream and unaccountable.'[37] As Guerrini observes, given Pitcairne's belief in hereditary right, 'the latter point was crucial'.[38] Thus in 1707, whilst still seeking patronage for his Newtonian protégés who were being overlooked for professorial posts at St Andrews and Aberdeen, a frustrated Pitcairne was privately telling Lord Mar that 'Medicine belongs to Mathematics, and Divinitie to South-britain'.[39]

Throughout the 1690s Pitcairne's correspondence reveals his keen interest in the theological implications of Newtonianism, as he repeatedly urges David Gregory at Oxford, 'to procur me a scheme of Mr. Neuton's [*sic*] divine thoughts, (I'l hope yee'l not laugh), that I may write a demonstration for our religion'.[40] Pitcairne planned to write a 'Relligio mathematici, or Euclidis' and 'ane immortal confutation of poperie' which would be so controversial that it would have to be published posthumously.[41] He was clearly no Romanist, but in 1706 he urged Gregory, 'For God's sake keep Sir Isaac at work, that wee may have … his thoughts about God … I am clear that metaphysics can never prove a Deity, and therfor [*sic*] think our churchmen here have no ground not to be Atheists; but for this, get a sight of Cromarties book … '.[42] Pitcairne refers to *A Right Use of Reason Against Atheists and Deists* (1705), by George Mackenzie, Earl of Cromartie (1630–1714), who (just like 'Salathiel') employs tortured scholastic Aristotelian terms to argue that religion 'stands on a surer Word of prophecy, than what Sense can perceive, or [fallen] human Reason can argue, or Flesh and Blood, i.e. humane Nature, can discover'.[43] Pitcairne summarily dismissed such rearguard defences of revelation and their a priori metaphysical assumptions – not, it would seem, as an atheist, but because such outmoded arguments undermine academic authority. He was almost certainly privy to the fact that Newton privately entertained heretical theological positions. But, although Pitcairne also privately expressed scepticism regarding the established metaphysical proofs for revelation, whatever his personal doubts, his medical essays assume a deity as the ultimate basis of a fixed natural and social order. He did, however, want that order to be Episcopalian and hierarchical, not Presbyterian and democratic.

This mixture of conservatism and modernity is evident in Pitcairne's 1688 essay *Solutio Problematis de Historicis* ('The Solution of the Problem Concerning Inventors'). In this defence of Gabriel Harvey's priority claims regarding the discovery of the circulation of the blood, Pitcairne is a Modern, arguing against those Ancients who claimed that Hippocrates had anticipated

Harvey. Although Pitcairne opens by denouncing 'the Credulity of the People, and what naturally follows that, a perpetual Desire for Innovation', in effect he argues for a new concept of intellectual copyright.[44] He also presents a rigorous critical methodology for establishing 'Demonstrative Certainty' and verifying historical evidence by distinguishing between 'Those things which are demonstrated by their own Evidence, and those that are so by the Light of other Things'; an example of the first being 'The Whole is greater than the Part' (demonstration), and of the second, that 'Phythagoras found out the 47th Proposition of the first Book of Euclid' (only a probability). We can only be absolutely certain of things demonstrated because those merely taken on credit from history, which have been passed down verbally over generations, are open to distortion. In short 'an Argument drawn from the Credit of such Histories, is not of Force against Demonstration and the Evidences of Sense'.[45]

When Pitcairne opens his *Solutio* with the declaration that 'I shall use the Terms of Author, Inventor, Observer, and Historian, for one and the same Idea' he almost encourages the application of such empirical methods to Scriptural authority; an implication which did not pass unnoticed.[46] Appended to *Apollo Mathematicus* (1695), a pamphlet by rival satirist-physician Edward Eizat there is a parodic 'Discourse of Certainty … ', which ridicules Pitcairne's historical scepticism in the *Solutio* by suggesting that upon his grounds of proof we would never be able to say if there had ever been 'a Charles IInd' or know if 'a man was born of woman … or grew out of the Ground like a Cabbage'.[47] Eizat finds Pitcairne's essay an 'impudent piece of Stuff' which 'shakes the Foundation of all historical Certainty, whether the History be Sacred or Profane'.[48] This leads to empiricism in medicine and 'tends to Scepticism' in theology by raising epistemological issues which threaten the very 'root' of revealed religion: 'And as he that regulats [*sic*] his Practice of Physick by the first, will at best prove nothing but a Mountebank; so he that founds his belief on the last, will prove little better than a Deist or Infidel.'[49] Eizat generously accuses Pitcairne of 'sublime Inadvertency' rather than atheism, but like Wodrow charges him with contributing to a general movement towards scepticism:

> There is a Set of Men sprung up, who call themselves *Deists* (but who for the most part are down-right *Atheists*), these pretend to be the only Men of Logick and Disputation, who make the learned Mob stand amazed at the wit and profoundness of their idle Chat, as if they had seen the Gorgon's Head …. These Wits impudently tell us, that nothing is certain but a Mathematical Demonstration, and that all historical certainty amounts to no more but meer Conjecture, and that the best attested History is little better than Romance: Tho' the best reasons they bring to maintain this Paradox are little better than dreams of Enthusiasts, or ravings of Sick Folks.[50]

This captures the tone of the debate in which, ironically, Pitcairne is accused of learned demagoguery. Eizat asks defensively: 'Have not these that have writ of the truth of the Christian religion, as clearly demonstrat [*sic*] that there was such a one as CHRIST JESUS, who was Crucified as Jerusalem, rose again from the Dead etc. as any thing can be: Matters of fact that the greatest Adversaries to the Truth never had the impudence to deny?'[51] Intentionally so or not, the *Solutio* could be read as casting doubt over the historical evidence for the supernatural basis of Christ's existence.

Battle of the books

There is no evidence for Pitcairne ever being formally charged with blasphemy, but on at least one occasion he did feel obliged to defend himself before the Edinburgh judiciary when one 'Mr Webster had pretty publicly called Doctor Pitcairne an Atheist'.[52] This incident, as recorded in Lord Fountainhall's legal *Decisions*, occurred on 18 July 1712:

> In February last Mr James Webster, one of the Ministers of Edinburgh, and Mr Robert Freebairn bookseller there, being accidentally in company of the magistrates, Mr Webster complained that Freebairn in his auctions sold wicked and prohibited books, and particularly Philostratus's Life of Appolonius Tyanaeus, wherein that vile imposter and magician is equalised if not preferred to our blessed Saviour, and his miracles, and which are greedily bought up by atheists and deists and Mr Freebairn bidding him condescend, Mr Webster answered, Such as Dr Pitcairn, who is known to be a professed deist; and for a further proof of it, in that same auction, many striving for that infamous book, one regretted that there were none bidding the bible, the doctor scoffingly answered, that it was no wonder it stuck in their hands for *verbum Dei manet in aeternum* [God's word remains eternal]: which was a direct ridiculing the Scriptures, and the Christian religion. In the heat of this contest, Freebairn takes instruments on Mr Webster's words, and carrying the relations to the Doctor, he raises a process against the said Mr Webster, before the sheriffs of Edinburgh, for redress of that atrocious injury, as tending to make him odious to all who own God and revealed religion, and so ruin and break his employment.[53]

The intermediary here is Robert Freebairn, the Jacobite printer who had assisted Pitcairne's often devious manoeuvres over the pseudonymous printing of Fever Dispute pamphlets, and who subsequently collaborated with Pitcairne's protégé, the literary editor Thomas Ruddiman.[54] Elsewhere it is recorded that after the contretemps at the book auction, Webster sought help from a 'shrew-witted' friend Mr Pettigrew, a Govan minister, who approached the physician in the street 'and tapping him on the shoulder, said: "Are you Dr Pitcairne, the atheist?"', to which 'the doctor, in his haste, overlooking the latter part of the query, answered "Yes"'. Pettigrew then called upon passersby as witnesses. Outwitted, Pitcairne remarked 'Oh, Pettigrew, that skull of yours is as deep as hell' to which the minister replied that he was glad 'to find you have come to believe there is a hell'.[55] As a consequence Pitcairne allegedly abandoned his defamation charge against Webster.

The controversial text at the auction was an English translation of Philostratus's 'Life of Appolonius Of Tyana' published in 1680 by the deist, Charles Blount (1654–93), a disciple of Hobbes who had corresponded with Rochester. Appolonius was a contemporary of Jesus, a neo-Pythagorean sophist also credited with prophetic insight and the ability to perform miracles, including restoring the dead to life. Although Philostratus's biography (AD 217) defends Appolonius against charges of charlatanism, by the third century Appolonius had became a cult focus for pagan opponents of Christianity. Blount's aim in printing his translation had been to historicise Christ's life and show that he was just one of many contemporary sages who claimed to possess supernatural powers. When the Edinburgh physician Charles Oliphant claimed that his rival Pitcairne 'bantered with Scripture' by 'calling it a Syrophoenician [*sic*] story', it suggests that Pitcairne had discussed, if not promoted, Blount's historicism.[56] Indeed it seems likely that Pitcairne was responsible for Blount's titles being amongst other sceptical works in Edinburgh University Library in the 1690s, a fact Hunter notes with some surprise.[57]

The Webster contra Pitcairne incident belongs with earlier controversies over the availability of sceptical texts at Edinburgh during the 1690s, as described by Hunter. In 1696 the Privy Council was 'ordering "a kind of inquisition"' of booksellers' shops in Edinburgh for titles deemed 'Atheisticall, erroniouss or profane and vitious [*sic*].' Titles cited included those of Blount, of the physico-theologian Thomas Burnet, and of the early seventeenth-century Italian apostate Julius Caesar Vanini. As Hunter shows, this attempt to outlaw books was linked to the 1696 trial for heterodoxy of John Fraser, an Edinburgh merchant's apprentice, which precedes the famous Aikenhead case. Fraser was charged with reasoning against the being of a God, of denying the immortality of the soul and the existence of the Devil, and of ridiculing the divine origin of the Scriptures, which he affirmed 'were only to frighten folks and keep them in order'. Although Fraser had reputedly called himself an atheist, he recanted before the Privy Council, claiming that he had simply been repeating views that he had read in Blount's compendium *The Oracles of Reason* (1693). Fraser's case was the first to come to trial after the re-enactment in 1695 of the lapsed 1661 blasphemy laws; a revival which Hunter shows to be symptomatic of the anxious climate amongst Presbyterians during the 1690s. Although found guilty, Fraser was treated less harshly than Aikenhead: having given public satisfaction for his misdemeanour in sackcloth, he was released on 25 February 1697.[58]

Pitcairne's personal library provided an unregulated source of potentially subversive texts which rivalled that of the university. It had grown to more than 1,500 volumes when it was eventually sold on behalf of Pitcairne's widow to Tsar Peter the Great (it survives in the Russian Academy of Science, St Petersburg).[59] Pitcairne's correspondence reveals a pattern in his bookbuying as he targets a number of proscribed titles. Possibly prompted by personal contacts in Florence and Padua, Pitcairne was keen to obtain the writings of humanist heretics of the Italian Renaissance. In 1702, for example, he had 'a singular use for' the works of Giordano Bruno who, with his writings condemned earlier by the Inquisition, had suffered a gruesome public execution at Toulouse in 1619. In 1694 at the sale of the library of Dr Andrew Balfour, Pitcairne was eager to gain a work which would be specifically condemned by the Edinburgh Privy Council two years later, Vanini's *Dialogi de Admirandis Naturae Arcanis* (1612), and gift it to his associate Dr John Gray. Vanini was one of the major disseminators of philosophical scepticism and atheism in early seventeenth-century Italy.

With such proscribed texts in circulation it is unsurprising that Pitcairne's circle were branded freethinkers and rational religionists. On 5 May 1710, the Presbyterian apologist Thomas Halyburton DD (1674–1712), devoted his inaugural oration, as newly appointed Professor of Divinity at St Mary's College, St Andrews, to a condemnation of Pitcairne as the chief promoter of the rampant deism polluting the religious life of Scotland.[60] As Emerson notes, Halyburton spoke with all the eagerness of the convert for, as recalled in his own memoirs, he had himself undergone a crisis of faith whilst 'engaged in the study of metaphysics and natural theology' at St Andrews in the mid-1690s: being 'accustomed to subtile notions, and tickled with them ... satan, in conjunction with the natural atheism of my heart, took occasion to cast me into racking disquietude about the great truths of religion, more especially the being of God.'[61] Unaware that he might be simply encouraging curiosity, the reformed Halyburton later appends to his *Natural Religion Insufficient* (1714), a list of books promoting deism and atheism, which includes titles by Pitcairne alongside 'Aikenhead's Speech' (made at his hanging).

Conclusion

The collective evidence for Pitcairne's reputation as a sceptic does not reveal the sustained statements of philosophical atheism attributable to Blount or his unfortunate readers Frazer and Aikenhead, though the physician does appear to have privately understood that mechanistic explanations posed a threat to traditional metaphysics. Whilst deliberately antagonising the Presbyterian establishment, Pitcairne's claims for mathematical infallibility suggested to some a certain ungodly arrogance. His empirical approach to the emergent issue of scientific priority in turn raised epistemological questions concerning the reliability of all historical evidence and hence of scriptural authority. More generally, his controversial and complex career reveals how, at the close of the seventeenth century, natural knowledge was not a coherent and unified discourse discrete from its social and political context of production, but rather a body of rival, variously powerful truth-claims co-existing dynamically within the context of sporadic, sometimes rather arbitrary, state control. Such claims were not without their social consequences. For a few, who argued too openly that the natural world was comprehensible without recourse to God as either architect or providential controller, those consequences could be fatal.

Notes

1. See Editors' Introduction above. For Edinburgh medical culture, see Riley, P.W.J. (1979), *King William and the Scottish Politicians*, Edinburgh: John Donald; and Dingwall, H.M. (1995), *Physicians, Surgeons and Apothecaries: Medicine in Seventeenth-Century Edinburgh*. East Linton: Tuckwell Press, esp. pp. 8–12.
2. Guerrini, Anita (1993), '"A Club of Little Villains": rhetoric, professional identity and medical pamphlet wars', in M. Roberts et al. (eds), *Literature and Medicine During the Eighteenth Century*, London and New York: Routledge, pp. 226–44.
3. The 'e' is often absent, but 'Pitcairne' is the spelling of his signature. Johnston, W.T. (ed.) (1979), *The Best of our Owne: The Letters of Archibald Pitcairne, 1652–1713*, Edinburgh: Saorsa; Webster, C. (ed.) (1781), *An Account of the Life and Writings of the Celebrated Dr Pitcairne*, Edinburgh and London: Gordon and Murray, Richardson and Urquhart; Thin, R. (1928), 'Archibald Pitcairne', *Edinburgh Medical Journal*, July, pp. 368–82.
4. Ritchie, Robert Peel (1899), *The Early Days of the Royal College of Phisitians Edinburgh*, Edinburgh: Johnston, pp. 159–87; Cunningham, A. (1981), 'Sydenham versus Newton: the Edinburgh Fever Dispute in the 1690s', in W.F. Bynum and V. Nutton (eds) *Theories of Fever from Antiquity to the Enlightenment*, London: Wellcome Institute.
5. Guerrini, Anita (2000), *Obesity and Depression in the Enlightenment: The Life and Times of George Cheyne*, Norman: Oklahoma University Press, p. 30.
6. Habermas, J. (1989), *The Structural Transformation of the Public Sphere*, trans. T. Burger and F. Lawrence, Cambridge: Cambridge University Press.
7. *Spectator* (1711), no. 10; Duncan (1987), p. 57.
8. Duncan, D. (1987), 'Scholarship and Politeness in the Early Eighteenth Century', in A. Hooke (ed.), *The History of Scottish Literature (1660–1800)*, gen. ed. C. Craig, 4 vols, Aberdeen: Aberdeen University Press, vol. 2, pp. 51–63.
9. Emerson, R.L. (1988a), 'Science and the Origins and Concerns of the Scottish Enlightenment', *The History of Science*, **26**, 333–65; (1988b) 'Sir Robert Sibbald, Kt, the Royal Society of Scotland and the Origins of the Scottish Enlightenment', *Annals of Science*, **45**, 41–72.
10. Emerson, R.L. (1989), 'The religious, the secular and the worldly: Scotland

1680–1800', in J. Crimmins (ed.) *Religion, Secularization and Political Thought*, London: Routledge, pp. 68–89 (p. 73); Cunningham, A. (1977), 'Sir Robert Sibbald and Medical Education, Edinburgh, 1706', *Clio Medica*, **13** (2), 135–61.

11. Macqueen, John (1982), *Progress and Poetry: The Enlightenment and Scottish Literature*, Edinburgh: Scottish Academic Press, p. 4.
12. Duncan (1987), pp. 51–2.
13. Emerson (1989), p. 73.
14. Bower, A. (1817), *The History of the University of Edinburgh*, 2 vols, Edinburgh and London: Oliphant, Waugh, Innes and Murray, vol. 1, p. 427; Ouston, H. (1982), 'York in Edinburgh, James VII and the Patronage of Learning in Scotland 1676–1688', in J. Dwyer et al. (eds), *New Perspectives on the Politics and Culture of Early Modern Scotland*, Edinburgh: John Donald, pp. 133–55.
15. Hannay, R.K. [n.d.], 'The Visitation of the College of Edinburgh in 1690', *Book of the Old Edinburgh Club*, **8**, 79–100.
16. Lindenboom, G.A. (1963), 'Pitcairne's Leyden Interlude Described from the Documents', *Annals of Science*, **19**, 273–84.
17. Guerrini, A. (1987) 'Archibald Pitcairne and Newtonian Medicine', *Medical History* **31**, 70–93; (1989), 'The Tory Newtonians: Gregory, Pitcairne and their Circle', *Journal of British Studies* **25**, 288–311; Shirlaw, L. (1975), 'Dr Archibald Pitcairne and Sir Isaac Newton's "Black Years" (1692–94)', *The Chronicle of the Royal College of Physicians of Edinburgh*, pp. 23–6; for anatomy, see Thin (1928), p. 379, and Dingwall (1995), pp. 72–3.
18. Phillipson, N. (1989), *Hume*, London: Weidenfield and Nicholson, p. 146, n. 8.
19. Hunter, M. (1992), '"Aikenhead the Atheist": The Context and Consequences of Articulate Irreligion in the Late Seventeenth Century', in M. Hunter and D. Wooton (eds), *Atheism from the Reformation to the Enlightenment*. Oxford: Clarendon Press, pp. 221–54.
20. Ibid., p. 239.
21. Ibid., p. 239–40.
22. Pitcairne probably had private medical pupils; see Finlayson, C.P. (1953), 'Two Highland Protégés of Dr Archibald Pitcairne', *Edinburgh Medical Journal* **60** (3), 52–60.
23. From the anonymous preface to Pitcairne A. (1830), *Babell: a Satirical Poem on the Proceedings of the General Assembly in the Year M.DC.XCII*, Edinburgh: Maitland Club, p. xii.
24. For Bower, see Johnston (1979), pp. 45 and 52.
25. Pitcairne (1830), pp. vii–viii.
26. Wodrow, R. (1843), *Analecta or Materials for a History of Remarkable Providences mostly relating to Scotch Ministers and Christians*, 4 vols, Edinburgh: Maitland Club, vol. 2, pp. iv–vii.
27. Burnet, G. (1680), *Some Passages in the Life and Death of John Wilmot, Earl of Rochester*, London.
28. Recorded in Watkins, J. (1808), *Characteristic Anecdotes of Men of Learning and Genius*, London: Cundee (p. 319), citing a possible 'source' in the writings of Ceasar Baronius.
29. 'Ad Greppam', in Pitcairne A. (1727), *Selecta Poemata*, p. 16; Thin (1928), p. 368. 'Greppa' refers to the 'groping' required to enter.
30. 'Doctor Pitcairn's [*sic*] Fareweel to the Surgeon's Hall', 17 August, 1711, Edinburgh University Library, Laing MS, fol. 358, item 5.
31. Chambers, R. (1868), *Traditions of Edinburgh*, Edinburgh: Chambers, pp. 158–60.

32. Freeman, F.W. and Law A. (1979), 'Allan Ramsay's First Published Poem: the Poem to the Memory of Dr Archibald Pitcairne', *The Biblioteck*, **9**, 153–60.
33. Johnston (1979), pp. 20–21.
34. Guerrini (2000), p. 30.
35. Tobin, T. (ed.) (1972), *The Assembly by Archibald Pitcairne*, Lafayette, IN: Purdue University Studies, pp. 82–3.
36. Ibid., p. 65.
37. Idem.
38. Guerrini (2000), pp. 27–8.
39. Johnston (1979), p. 45.
40. Ibid., p. 19.
41. Ibid., pp. 18–19.
42. Ibid., p. 43.
43. Mackenzie, G. (1705?) *A Bundle of Papers, Partly Self-Evident, partly Problematick raised from Occasional Meditations*, National Library of Scotland, p. 1.
44. Pitcairne A. (1715), *The Works*, trans. G. Sewell and J.T. Desaguliers, London: Curll, p. 139.
45. Ibid., p. 143.
46. Ibid., p. 141.
47. [Edward Eizat] (1695), *Apollo Mathematicus or the art of curing Diseases by the Mathematicks according to the principles of Dr Pitcairne*, London: n.p., pp. 15–16.
48. Ibid., p. 13.
49. Ibid., pp. 3–4.
50. Ibid., p. 4.
51. Ibid., pp. 14–15.
52. Wodrow (1843), p. 307.
53. Lauder of Fountainhall, J. (1759–60), *Decisions of the Lords of Council and Session*, Edinburgh: n.p., vol. 2, pp. 756–7.
54. Duncan, D. (1965), *Thomas Ruddiman: a Study in Scottish Scholarship of the early Eighteenth-Century*, London: Oliver and Boyd.
55. Chambers (1868), p. 60.
56. Oliphant, C. (1702), *A Short Answer to Two Lybels Lately Published*, Edinburgh: n.p.
57. Hunter (1992), p. 248; Pitcairne donated books to the RCPE Library, the University Library and the Advocate's Library.
58. Hunter (1992), pp. 239–42.
59. Appleby, J.H. (1986), 'Archibald Pitcairne Re-Encountered: A Note on his Manuscript Poems and Printed Library Catalogue', *The Biblioteck*, **12** (6), 137–9.
60. Halyburton, T. (1714), *Natural Religion Insufficient, Edinburgh and Oratio Inauguralis Habita Andreapoli* [1710] reprinted in his 1835 *Works*, London.
61. 'The Memoirs of ...' in Halyburton (1835), p. 701.

PART II
RELIGION, POLITICS AND THE NATURAL WORLD

CHAPTER FIVE

The Politics of Morbidity: Plague Symbolism in Martyrdom and Medical Anatomy

Peter Mitchell

The narrative semiotics of plague in images of Sebastian

In England, France, Italy and Spain medical treatises on plague, or pestilence, ascribed its aetiology to a combination of contagion, miasma and 'the wil of God rightfully punishing wycked menne'.[1] The theory of miasma located plague's aetiology in the corrupt and infectious air which arose from, among other things, decaying corpses.[2] It supplemented rather than challenged divine aetiologies, because it tended to be accompanied by an ethically-informed pathology, in which the disease resulted from a corruption of the humours and a dissipation of vital spirit. These owed their provenance to 'the abuse of thinges not naturall', principally involving in the first case immoderate appetite, emotion or rest, and in the second sexual excess.[3] In Italy, initially, and later elsewhere, the contagion theory did encourage public health measures, but a divine aetiology was fuelled by the virulence of the disease.[4] And for the Counter-Reformation Church, the invocation and intercession of saints constituted a symbolic and thaumaturgical solution to the medical and social crisis of epidemics.

Whereas the efficacy of other 'plague saints' was frequently confined to a city,[5] Saint Sebastian was given the title *Defensor Ecclesiae*, by Pope Caius, as defender of the members of the 'one body' of Christ from both disease and heresy. The twenty-fifth session of the Council of Trent (1545) further defined as a 'legitimate use' of saints' images, the invocation of those saints. Consequently images of Sebastian were commissioned in response to outbreaks of plague, by churches and hospitals, where the saint could be invoked to defend the afflicted and faithful. Intercession could be artistically represented as the saint shielding the faithful from God's arrows of punishment. Although arrows were a metaphor for disease in general, they were closely linked with the heavenly and miasmatic aetiologies of plague, from which deliverance was sought.[6] Intercession, however, more often involved the saint's body, besieged by arrows, suffering on behalf of those desiring recovery or wishing to avert infection. Jean ver Meulen articulated one systematic entailment of this metaphor of arrows for plague in his *De Picturis et Imaginibus Sacris* (1570): 'the arrow wounds call to God for mercy for us, as the symptoms of the infirm call for pity from the passerby'.[7] If the entry of the arrows into the saint's body symbolises infection, then the wounds become the symptoms, the buboes and carbuncles.[8] The symptomatology of plague, and intercessory drama, thereby overlaid the martyrdom of the legend.

Furthermore artists may have overlaid the legend of Sebastian's martyrdom with a narrative of disease beginning with infection and ending in recovery.[9] Epidemics in Europe in the late sixteenth and seventeenth centuries spurred the production in Catholic cultures of images of Sebastian's cure by Irene, which in the legend follows his sagittation.[10] The Counter-Reformation Church had turned to this episode of the legend to legitimise Sebastian's representation in images, in the face of accusations of idolatry, impiety and the inclusion of apocryphal or extraneous episodes in sacred paintings. But rather than merely assuming the ostensible veracity of the legend, the cure, whether by Irene or (as a variant) by angels, was overlaid with the symbolism of plague treatment.[11]

Of the paintings of Sebastian assisted by Irene, one contains a peculiar conceit which seems to combine the martyr's sagittation with his cure. This seventeenth-century Italian painting is a copy of a work by Francesco Cairo, and has been attributed to Michelangelo Merisi da Caravaggio (1571–1610), Bartolomeo Manfredi (1580/2–1620/1), and Orazio Riminaldi (1586–1630) (Figure 5.1). It depicts Irene anointing Sebastian's wounds with a feather. Given that there is no necessity for ointment to be applied using a feather, it is not the ointment that is the nodal point of the semiotics of the painting, but the startling presence in the therapeutic procedure of a part of the instrument responsible for injuring the saint's body. The feather seems to function as a synecdoche for the whole arrow, which appears in the painting close to the site of Irene's application. And the whole arrow metonymically signifies the inferred sagittation, and the divine and miasmatic aetiologies of plague. Within the pictorial space, and along the narrative of the martyrdom, a metonymic chain is created between sagittation, which punitively injures, and feather, which heals. Since the persuasive power of Sebastian's intercession was believed to lie in his immense suffering, it is conceivable that his sagittation was regarded as directly analogous to the therapeutic process, whereby the Catholic patients had received (or so it was thought) their cure.

Sebastian appeared to reconcile the charity of intercessory healing with the salutary example of his martyrdom. That the sagittation and the cure were

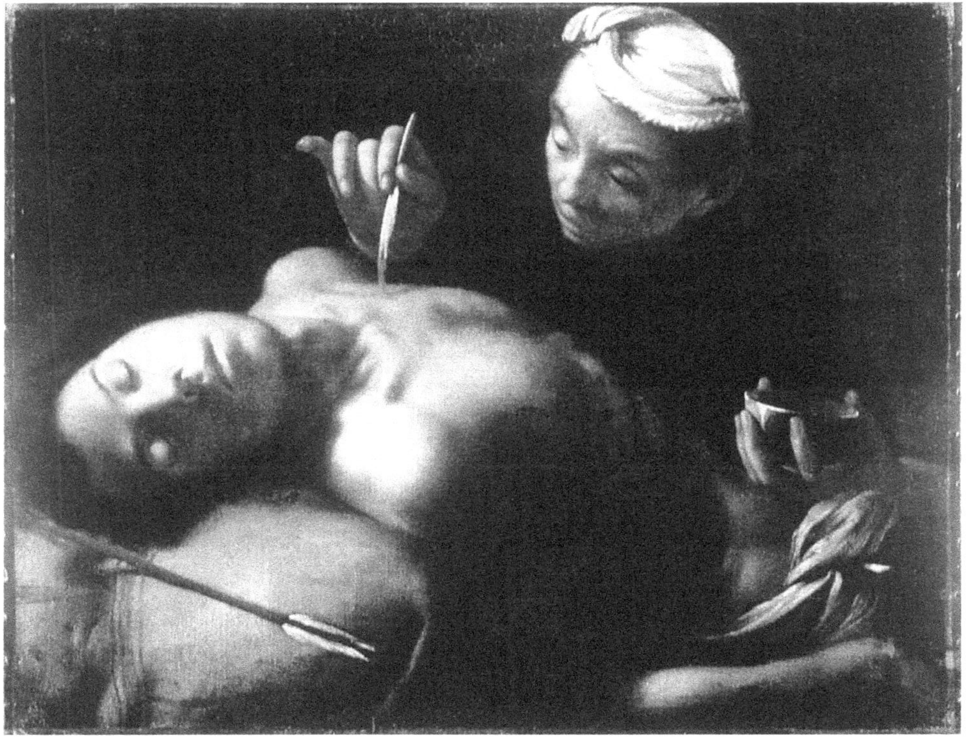

5.1 Anon., *San Sebastiano Curato da Irene (Saint Sebastian Assisted by Irene)*, oil on canvas, 17th century.

perennially elected by artists, is partly because they were amenable, respectively, to asceticism ('a complete resignation to the holy will of God')[12] and medicine. The analogy between the recoveries of Sebastian and of plague sufferers lends authority in the figure of Irene to medical intervention in a disease regarded as God's punishment, and thereby risks ignoring the Florentine chronicler Giovanni Cambi's warning, that 'further infection was due to God's punishment for Man's presumption for thinking that he could do something about the spread of the pest'.[13] A consideration, though, even for the staunchly divine aetiologist Henoch Clapham, is that 'to neglect such naturall means as reason and experience have found out to availe against naturall infirmitie [*Deo non obstante*] the Lord not crossing nature', is to leave oneself open to the accusation of tempting God to demonstrate His love and omnipotence.[14] Nevertheless, there is a contradiction rooted in an artistically rendered religious paradox. Although the logical consequence of martyrdom is death, Sebastian is painted frequently as having gained distance from this mortal threat, not through his cure, but through either stoical composure, reflective sadness or mystical ecstasy.

The Council of Trent confirmed the legitimacy of the belief that martyrs, displayed in sacred images, should be emulated, in order 'to adore and love God, and to cultivate piety'.[15] However, the Council did not set limits on such emulation, which easily tips over into an appropriation of the agency of persecution or punishment, by the ascetic, with the intention of producing a salutary effect on the body and soul. Penitential self-torture, in forms such as self-inflicted bleeding, gave morbid expression to the fallen condition 'of putrefaction on earth so as better to breathe heavenly aromas'.[16] By contrast, medicine implied not merely the existence, but the desirability, of the converse state of health on earth. The conflict of models was acutely felt in the Counter-Reformation, when the Roman Church suppressed, mitigated or gave 'more effective form and more refined manner' to medieval ascetic practices. The 'hyper-asceticism of the Middle Ages' was disapproved of by the Reformation churches,[17] but even beyond the emulation of saints, some Protestant voices could also be heard recommending gratitude for the 'lenitie and fatherlie kindnes' of plague's swift punishment.[18] Exposed by the contradiction which might thus plausibly have emerged in the Catholic beholder's experience of paintings of martyrdom, and in any Christian's perception of disease, or more broadly morbidity, as the will of God, is an almost intractable difference between asceticism and medicine.

In representing Sebastian's sagittation as though it were his death, paintings of the martyrdom seem to have overcome this almost intractable difference. The ubiquitous centrality of the figure of Sebastian, his nudity, and his elevation on a pedestal, or on the truncated branches of a tree, all testify to the iconographic influence of the crucifixion upon the martyrdom. The martyred Sebastian's occasional proximity to death gestures towards a notion that, like Christ, he actually died from his wounds, and later resuscitated.[19] From such iconography and deathly countenance it was appropriate for the viewer piously to extrapolate that ideally all emulative injury and morbidity should be thus deathly, in imitation of Christ; and should, given the prohibition on suicide, be inflicted upon one as a punishment, whether for one's own or others' sins. Agency, having been appropriated by asceticism, is thereby reappropriated and contained as a willed response appropriate to affliction, torment or disease, categories corresponding to the broadest and oldest definitions of plague. The disease category to which plague more specifically refers, after about 1548–49,[20] was far from difficult to conceptualise as analogous to death. This conceptualisation was facilitated by its frequent prognosis,[21] and its theological and miasmatic aetiologies, but also by the degenerative and putrefying nature and symptomatology of the disease, as it was understood.[22]

The representation of Sebastian's recovery as bodily resurrection was advanced by the seventeenth-century vogue for depicting the episode with Irene. The iconography and spatial arrangements of treatments of this theme parallel those of *pietàs*, depositions and lamentations.[23] In José (Jusepe) de Ribera's *San Sebastián Curado por las Santas Mujeres (Saint Sebastian Cured by the Holy Women)* (1621), the martyr's slumped horizontal body and still-tied wrist suggest a *deposition*; while the wound, from which Irene extracts an arrow, as the subject of her intent gaze, draws upon the motif of the Doubting Thomas legend. This motif reoccurs in a painting attributed to a follower of Ribera, wherein the gaze of Irene is directed towards 'the hand of the servant which is touching the wound in Sebastian's side'.[24] The inquiry into Sebastian's wound, equally pious and searching, is into the saint's survival of his sagittation, but also, in the doctrinal tradition of the wounds of the risen Christ, into the truth of his 'resurrection'. The wounds or 'red letters' of the risen Christ were interpreted, as Richard Crashaw reads them in his divine epigram 'On the Still Surviving Marks',[25] both as traces of how Christ's corporeal body was killed and as evidence that His risen, transformed body is demonstrably His own, and, hence, as assurances of the resurrection. Encouraging the similarities between Christ and Sebastian was the view that the wounds of martyrs constituted a true imitation of Christ. Envisaging the wounds as letters or words invites the metaphorical entailment that the nails or the arrows are writing instruments,[26] thereby comparing the instruments of the passion or martyrdom with those of literary creation. Similarly, the pictorial skill displayed in the variously attributed painting of Sebastian assisted by Irene (see Figure 5.1), draws attention to itself by the visual analogy between the use of the feather and the artist's brush. These touches of self-conscious artistry propose the efficacy of the eloquent poem or ingenious image, to inspire adoration of God and faith in His promise of salvation or deliverance.

The eucharistic medicine of Pacheco's *San Sebastián Asistido por Santa Irene (Saint Sebastian Attended by Saint Irene)* (1616)

One painting of exceptional iconography uses sacramentalism to represent the medical efficacy of Sebastian's ascetic sacrifice. In 1616, the Sevillian artist Francisco Pacheco (1564–1654) chose to represent the martyr recuperating in bed, in the house of Irene (Figure 5.2). A window in the wall affords the peruser of the painting a view of an agitated, dynamically mannerist treatment of the saint's sagittation. Pacheco was aware that for the faithful the intercessory and morbid significance of painting Sebastian 'full of arrows and wounds, is that he is showing each wound and calming us through so many mouths, asking mercy for us; and he frees us from contagious evil'.[27] The Renaissance, rather than medieval, version of the artistic convention of continuous or polyscenic narrative is employed,[28] to signify the saint's composure after surviving his ordeal.

A tentative conversion of the painting's signification to a sacramental register occurs on account of the paten-like rendering of the saucer or shallow dish, on which rests a bowl of 'pink syrup'.[29] The salutary effects of the eucharistic body and blood were thought by seventeenth-century Jesuit and other theologians to extend to the recipient's body.[30] While Sebastian seems thereby to represent a recipient, additional sacramental iconography congregates to suggest that his regenerated body is itself in some sense the sacrament. If so, Sebastian's position above the altar-like bed, covered with linen, is like that of the host itself, and seems also to allude to the notion that the glorified bodies of the martyred saints will be 'above the altar' at the Last Judgement.[31] Through the painting's equivocal features, another narrative has emerged, in which the sagittation signifies a death comparable with that of Christ, as a result of which the redeemed martyr's incorporeal risen body somehow becomes a sacrament.

Pacheco's painting converts the naturalistic, spatialised narrative of the legend to a sacramentalism consistent with Counter-Reformation theology in respect of the sacrifice and real presence of the eucharist.[32] From the conversion there emerges a correspondence between two sets of relations: between the real sacrificial martyrdom and the sacrament; and between the saint's arrow-infested corporeal body and his body above the altar. The arrangement suggests that the sacrament of the body and blood of Sebastian does not contain the real martyrdom, but signifies it (through the window). But it does contain the saint's body, risen above the altar, accompanied by the

5.2 Francisco Pacheco, *San Sebastián Asistido por Santa Irene (Saint Sebastian Attended by Saint Irene)*, oil on canvas, 1616.

accident of pink syrup. The incorporeal body's spatialised appearance in corporeal form on the bed, rather than under the accident of bread, functions figuratively, rather than naturalistically, to signify the real presence, and to assert a continuity of identity with the corporeal body through the window.

Furthermore, Pacheco's painting reconciles an artistic decorum imbibed from the Council of Trent, which was to dissuade artists from any lasciviousness in their representations of the saints,[33] with the painting's eschatological and sacramental significance. In the house of Irene, Sebastian wears a tunic, which is atypical of his images, and which covers wounds indicated by the actions of the maid, bringing 'a little glass of balm and some lint (for dressing wounds) on a plate'.[34] The tunic prevents the visibility customarily associated with the wounds that might otherwise act as metonymic traces of the martyrdom. This may display the artist's hesitation with regard to whether the resurrected bodies of martyrs would contain their wounds. But the lack of visual certainty is also consistent with the Counter-Reformation theory of the eucharistic sacrifice, in that the sacramental body ambiguously contains the signs of the sacrificial martyrdom.

The sacramental character of Pacheco's painting enforces one to ask how, and by whom, the sacrament is received. Sebastian may, as a martyr, be the recipient of the grace of his own sacrifice.[35] Equally, the sacrament may be imagined as being received by the patients of the hospital of the parish of Saint Sebastian, in Alcalá de Guadaira, for which the painting was commissioned as an altarpiece. If so, the patients are optimistically reflected by Sebastian recovering in the house of Irene, even as they receive his body as a sacrament. But how does the represented efficacy become or otherwise relate to the real efficacy for plague sufferers?

Eucharistic theology was, according to Regina Stefaniak, art theoretical in its eloquent terminology, and was consequently 'insufficient to define and distinguish a eucharistic body of Christ from a painted body of Christ'.[36] Nevertheless, the Counter-Reformation Church insisted on a distinction, and, as a Familiar of the Inquisition, whose duty it was, in his own words, 'to denounce the errors committed in pictures, by the ignorance or wickedness of the artist', it is unlikely that Pacheco would have risked contravening the canons and decrees of the Council of Trent over idolatry of images.[37] Rather, once the meaning of the painting is accepted as sacramental, the proposition emerges that a morbid condition participates in and benefits from the sacrifice of martyrdom. It does so through a sacramental medical efficacy, which may parallel the efficacy of Christ's eucharistic intercession as described by Thomas de Vio Cardinal Cajetan and other Counter-Reformation theologians.[38] If so, the sacramental intercession makes the recipients of the sacrament partakers of the remission of sins and deliverance from pestilence wrought by Sebastian's historical or legendary martyrdom. However, the relation between sacrament and historical sacrifice is ambiguous. The mannered exaggeration of the figure of Sebastian through the window in the painting displaces the unadmitted torment of his martyrdom.[39] Yet, the indignation of the recovering saint's expression in the painting is indicative of his struggle, and that of the Counter-Reformation, to accommodate medical recovery to the ascetic rigours of martyrdom.

While superficially depicting two episodes from the legend, Pacheco's densely symbolic canvas transforms morbidity and the recovery of health with concepts from eschatology and eucharistic theology; it appropriates them for theology, albeit at the indignation of asceticism. The methods it uses are deducible in a highly theorised form, because the artist was unusually imbued with the theological potentialities of artistic representation. But they were already employed outside of devotional art, in the reception, and possibly the production, of anatomical illustration.

Redeeming incision: the martyrisation of anatomical cadavers

Charles Estienne's *De Dissectione Partium Corporis Humani*, an illustrated anatomical textbook including a long series of male figures, was assembled between 1521 and 1545 and published in Paris in the latter year.[40] Among these illustrations are those described by C.E. Kellett as 'morbid and remarkable drawings'; their figures are 'hooked up on trees and propped up against crumbling masonry',[41] in ways reminiscent of Sebastian in paintings of his sagittation.[42] In these plates are a number of iconographic features of the artistic tradition of Sebastian's sagittation, for which the most prolific period falls earlier than those paintings which have featured prominently so far in this chapter, and culminates shortly after the production of the plates. It is a tradition, however, upon which Pacheco drew, in 1616, for the sagittation scene in his *Saint Sebastian Attended by Saint Irene*. The remainder of this chapter interprets the Sebastian plates doubly retrospectively: firstly, and inevitably, in the context of the present; secondly, from the perspective of the early seventeenth century. Thus, in this chapter, the historicity in question for the Sebastian plates is that of their 'inscribeability'[43] into early seventeenth-century Counter-Reformation culture. This involves that historical formation's terms of acceptance of the plates.

The figures on pages 168 (Figure 5.3), 180, 202 and 229 of *De Dissectione* appear against trees, sharing this motif with Sebastian in the sagittation scene of Pacheco's painting, and with a great number of martyrdoms diverse in conception, execution and period of composition. This, among other iconographic traditions of the martyrdom, continued into the seventeenth century. The anatomical figure on page 202 pushes the sole of his foot against

5.3 Charles Estienne, *De Dissectione Partium Corporis Humani libri tres*, Paris, 1545, p. 168.

a tree trunk in a variation of the traditional Flemish Renaissance pose for the martyr. Others, such as those on pages 168 (Figure 5.3), 180, 208 and 230, cast their eyes heavenward, as if surrendering to God's will or imploring Him to intercede in their suffering, a gesture which was 'an almost essential motif in the tradition of the isolated figure of the saint',[44] but occurred also in many paintings that incorporate the martyr's persecutors.

Iconographic indebtedness on this scale is unlikely to be adventitious. The exchange between art and anatomy has been extensively explored. Artists could intensify the emotional pitch of martyrdoms by mastering anatomy,[45] and the nude, often centrally placed, figure of Sebastian invited anatomical rendering, thus enabling anatomical illustration to draw from it a precedent for the relief, perspective and design of the surface anatomy it required. Such techniques, however, are acculturated, as Bettina Mathes discusses in Chapter 11 in relation to sexuality. And the assumption that what 'had to be provided to turn a martyrdom into an anatomy was a change in emotional climate'[46] is herein challenged in so far as it implies that the cathexis of plague sufferers was irrelevant to the plague symbolism of the Sebastian plates of *De Dissectione*.

One of three complex reasons provided by Jonathan Sawday for the occurrence in sixteenth- and seventeenth-century anatomical illustration of animated dissected figures in densely semiotic landscapes – 'living self-anatomy' – is to 'gesture towards that established tradition of admonition which is summarised in the motto *et in arcadia ego* – "Even in Arcadia, there am I, Death"'[47]. Panofsky's article 'Et in Arcadia Ego' pursues the same theme.[48] The illustrations thereby function as *memento mori*, and certain features permit this reading of the Sebastian plates of *De Dissectione*. The dissected body of the figure on page 168 leans wearily against a correspondingly blasted tree, one broken branch of which is only precariously attached to the main body of its life (see Figure 5.3).[49] Lingering exhaustedly on the edge of life, this, and the other figures in the Sebastian plates, conspire with their landscapes of crumbling masonry or storm-blasted trees, to be emphatic in their admonitions about mortality. However, the centrality of the figure on page 168 of *De Dissectione*, his partial elevation on the truncated branch of a tree, and one particular incision, point to the christological and eschatological emphasis of the iconography of mortality and transcience in paintings of Sebastian. Although the placement of incisions in anatomical illustration is subject to the area of anatomy intended for demonstration, a visual echo of Christ's spear wound occurs where the peritoneum of the figure is peeled back to reveal an incision low on the right side of the figure's abdominal wall. Notwithstanding coincidence as an explanation, the other christological iconography symbolises the aetiology, pathology and prognosis of plague, and its conceptualisation as analogous to death.

While the Sebastian plates inherit the iconography of sagittation, they apparently substitute dissection for it, and may on occasion literally do so, for several plates have their anatomical portions inset. Published as anatomical plates, they stage a production of anatomical meaning from an iconographic imbrication of martyrdom and medicine. In this production, dissection inherits the ambiguity of the sagittation, as both injurious and salutary. The reason for the representation of dissection as injurious, perhaps even mortal, and certainly divinely disciplinary, is complex. The constitution of a pestilent carbuncle is the necrosis of tissue, giving rise to the formation of an 'Eschar' (scar), which on occasion falls away to reveal internal anatomy as though by dissection. The prognosis of acute necrosis is 'great danger or death of the Patient',[50] and, taken together with plague's punitive divine aetiology, is consequently fit for analogies with the judicial execution and dissection of criminals.

The provision of cadavers from the scaffold for dissection, a disputed historical issue, is the second of the three reasons which Sawday provides for the convention of 'living self-anatomy'. In Sawday's view, the anatomist's direct or symbolic implication in the exercise of judicial power, and unavoidable proximity to the infamy of the executioner, sat uncomfortably with his desire for the cultural inclusion and acceptability of his profession.[51] However, perhaps more important than infamy,[52] in Italy, whence the designs for the Sebastian plates may have come to France,[53] *il manigoldo* (the executioner) was expected to play a 'vicarious devil's role' in the symbolic drama of execution, in which the kneeling victim was imagined 'as a pitiable soul struggling for salvation'.[54] The Sebastian plates literalise a symbolic transformation of the criminal into a saint. Perhaps surprisingly, then, the ambiguity of the sagittation commended itself to the anatomical illustrator. But, by substituting the tools of dissection for the arrows, anatomy justifies its direct or symbolic involvement in the rites of judicial power, by the contribution it makes to medicine. This quite specifically echoes a legitimation made elsewhere in medical literature, despite the lack of direct medical application for many of the anatomical projects of the sixteenth century.[55] As salutary, dissection in these plates inherits the efficacy of Sebastian's intercession for the faithful, rather than its asceticism.

This justification alone does not in any way modify the agency of criminal execution, but one effect of the conversion of the criminal to a martyr is to imply a willingness to die as a subject of anatomical investigation. The transfer of this assurance that the victim wished to die to dissection, where it is all that remains of Sebastian's ascetic piety, is conceivably what anatomists were looking for, to answer anxieties over the provenance of cadavers. In the Sebastian plates, this surrender of will most clearly characterises those figures on pages 168 (Figure 5.3), 180, 202 and 229. These figures are distinguishable from those which give most extravagant expression to the ubiquitous characteristic of the convention of 'living self-anatomy', that, in the absence of the anatomist, the cadaver is depicted as conspiring with its own dissection.[56] Graphically self-demonstrating figures do occur among the Sebastian plates, on pages 190 and 230, but those figures which most clearly resemble Sebastian in his martyrdom do not, and (as martyrs) need not, demonstrate in such graphic fashion.

It has been proposed that the convention of self-demonstration derives from two of Christ's gestures: the partition of his symbolic body at the Last Supper, and the exposure of the wounds in his risen body. With regard to the latter, Sawday proposes that an analogy was constructed between Christ as 'the subject of a gaze whose end was the establishment of the truth of his own resurrection', and the self-demonstrating cadaver which establishes the truth of anatomy's enquiry into the human interior.[57] Examining the visual demonstrations, however, in relation to the symbolic narrative episode of the cure of Sebastian, as was possible by the early seventeenth century, gives rise to an even greater analogical systematicity than that proposed by Sawday. When incisions are graphically demonstrated as the result of dissection, their function is directly analogous to that of the wounds. They establish the truth of the resurrection of the executed criminal's body as the anatomical figure. Like the recuperation or resurrection of Sebastian, this anatomical resurrection can be interpreted as medically efficacious. The criminal body is represented in the Sebastian plates as martyrised and redeemed by dissection and the graphic arts for the use of medicine. As Ludwig Choulant has remarked, the anatomical details of the figures are frequently too 'small and indistinct' to be of any service to the anatomist or the physician.[58] Notwithstanding this, their medico-religious significance exceeds 'knowing oneself in order to revere the Creator', which is the limit placed variously upon anatomy's religious significance in some recent studies.[59]

Although the iconographic indebtedness of the self-demonstrating figures of the Sebastian plates is not to paintings of the cure, it may be argued that, after the eucharistic controversy of the Reformation, if not before, these plates were inscribed with a sacramentalism such as that found in Pacheco's *Saint Sebastian Attended by Saint Irene*. If such an inscription occured, the Sebastian plates came to represent dissection as having performed a confection on the bodies of criminals, whereby those bodies became the species under which lie the incorporeal bodies of saints. The plates moreover, like a sacrament, purportedly perpetuate the efficacy of criminal dissection. Sawday argues that although the convention of 'self-demonstration' draws upon Christ's own division of his symbolic body at the Last Supper, the object of anatomical demonstration is not redemption but knowledge.[60] On the contrary, the martyrisation of criminals represented the relation between anatomy and the practice of medicine. In this relation, anatomical knowledge is represented as resembling a redemptive and sacramental intercession, in defence of the members of *ecclesia* from disease and death; a representation which precisely answers Stengelius's claim in the first volume of *Historia Pestis* (Augsburg, 1614), that since 'the plague is the visitation of God's anger on the wicked … only the veneration of the saints can help'.[61] Interpreted in this way, the representation effectively manipulates the eschatological and sacramental understanding of illness, for the dignity and cultural establishment of anatomical 'science'.

Morbid anatomy and pathology precipitated a dimunition in the ethical component of plague's aetiology from the late sixteenth century. In addition, public health measures encouraged by the contagion theory of plague's infectiveness began, in Italy, to take precedence over the invocation of saints.[62] But this alteration in plague's meaning did not go unchallenged, in either Catholic or reformed contexts, and advocacy of divine aetiology and its implications for recovery was sometimes suppressed. In England, in 1603, Henoch Clapham was accused of claiming 'The plague not to be infectious, and that All that dyed of the plague were damned, as dying without faith';[63] for which he was imprisoned. But from this contestation of meaning, medicine's interest in the health and integrity of the corporeal body had emerged carrying a positive religious significance; a significance which had moderated the position of these advocates, as it had advanced the propriety of scientific ingenuity's claim to deliver the body from morbidity. Meanwhile, Counter-Reformation representation of martyrdom relinquished its hold on the understanding of morbidity, eliding its ascetic rigours to reinscribe it as milder, mystical and ritualised.

Notes

For their comments on earlier versions of this chapter I am grateful to Professor John Manning, of the University of Wales, Lampeter, and Dr Lawrence Normand, of Middlesex University, whose additional recommendations at a late stage were helpful and incisive. I would also like to thank Mair Davies for her generosity in translating correspondence, and Miyõung Parc, for her model of scholarly courage.

1. Phaire, T. (1993), *The Regiment of Life, whereunto is added a treatise of the Pestilence* (1553), transcribed and intr. A.V. Neale and H.R.E. Wallis, London: British Paediatric Association, f. Mv *v*. Also see: Henderson, J. (1988), 'Epidemics in Renaissance Florence: Medical Theory and Government Response', in N. Bulst and R. Delort (eds), *Maladies et Société (XIIe–XVIIIe siècles)*, Paris: Centre National de la Recherche Scientifique, pp. 165–86, esp. pp. 169–70.
2. Paré, A. (1678), *The Works of that Famous Chirurgeon Ambrose Parey*, trans. T. Johnson, London: Printed by Mary Clark for John Clark, Bk. XXII 'Of the

Plague', p. 491. First published as *Les Oeuvres de M. Ambroise Paré*, 1575, Paris; first English edn 1634, further edns 1649 and 1665. Also see Henderson (1988), p. 167.

3. Phaire (1993), f. Nii *r.*, f. Mv *v*. Also see Paré (1678), pp. 492, 497.
4. Kiple, K.F. (ed.) (1993), *The Cambridge World History of Human Disease*, Cambridge: Cambridge University Press, p. 630, col. 2; Henderson (1988), pp. 166–7 and *passim*.
5. See, for example, Calvi, G. (1988), 'The Florentine Plague of 1630–33: Social Behaviour and Symbolic Action', in Bulst and Delort (eds) (1988), pp. 327–36.
6. See Zupnick, I.L. (1958), 'Saint Sebastian in Art', PhD dissertation, Colombia University, p. 190; Zupnick, I.L. (1975), 'Saint Sebastian: The Vicissitudes of the Hero as Martyr', in N.T. Burns and C.J. Reagan (eds), *Concepts of the Hero in the Middle Ages and the Renaissance*, Albany: State University of New York Press, pp. 239–67; esp. p. 239.
7. Cited in Zupnick (1958), p. 122.
8. Bubonic plague occured when *Yersinia pestis*, a rodent disease, was communicated to humans through the bite of infected fleas. *Y. pestis* 'rapidly replicates at the site of the flea bite. This area can subsequently become necrotic, where dead tissue blackens to produce a carbuncle or necrotic pustule ...'. 'The lymphatic system attempts to drain the infection to the regional lymph nodes', each of which 'becomes engorged with blood and cellular debris, creating the grossly swollen bubo.' Kiple (ed.), 1993, p. 629, col. 2.
9. For an analysis of a symbolic narrative of illness and recovery, with which the governors and people of Florence ordered the events of the plague epidemic there in 1630–33, see Calvi (1988), pp. 328–39.
10. For an extensive study of the Irene episode as a subject for painting in the seventeenth century, when it was in vogue, see Carr, C.K. (1964), 'Saint Sebastian Attended by Irene: An Iconographic Study', MA thesis, Oberlin College.
11. See, for example, Zupnick (1958), p. 147; Zupnick (1975), p. 257.
12. Hastings, J. (ed.) (1909), *Encyclopaedia of Religion and Ethics*, 13 vols, Edinburgh: T. & T. Clark; New York: Charles Scribner's Sons, vol. 2, p. 73, col. 2.
13. 'Istorie di Giovanni Cambi, cittadino fiorentino' (1786), in Di San Luigi, I. (ed.), *Delizie degli eruditi toscani*, Florence: n.p.; cited in Henderson (1988), p. 181.
14. Clapham, H. (1603), *An Epistle Discovrsing vpon the Present Pestilence*, London: Printed by T.C. for the Widow Newbery, f. B2 *v.*; also see f. A4 *v*.
15. *The Canons and Decrees of the Council of Trent* (1848), trans. Revd J. Waterworth, London: C. Dolman, p. 235.
16. Camporesi, P. (1988), *The Incorruptible Flesh: Bodily Mutation and Mortification in Religion and Folklore*, trans. T. Croft-Murray, Latin texts trans. H. Elsom, Cambridge: Cambridge University Press, p. 44.
17. Hastings (ed.) (1909), vol. 2, p. 78, col. 2 – p. 79, col. 1.
18. Clapham (1603), fols C2 *v.*–C3 *r*.
19. Sigerist, H. (1936), 'The Historical Aspect of Art and Medicine', *Bulletin of the Institute of the History of Medicine*, **4**, 271–97 (p. 285).
20. See Clapham, H. (1604), *Henoch Clapham His Demaundes and Answeres Touching the Pestilence* [London and/or Middelburg: R. Schilders], f. A3 *v*. (*Note*: Pagination is in folio A–B, although not always marked; and in page nos from p. 9, which is also f. B *r.*); Barnhart, R.K. (ed.) (1988), *Dictionary of Etymology*, Edinburgh and New York: Chambers, p. 801, col. 1.
21. See Kiple (ed.) (1993), p. 629, col. 2.

22. See Paré (1678), p. 490; Phaire (1993), f. Miiii *v*.
23. See Carr (1964), pp. 37–9, 51, 58–9, 64, 92.
24. Ibid., pp. 64–5.
25. Crashaw, R. (1957), *Steps to the Temple: Sacred Poems* (1646), in L.C. Martin (ed.), *The Poems English, Latin and Greek of Richard Crashaw*, Oxford: Clarendon, p. 86.
26. See Cunnar, E.R. (1994), 'Opening the Religious Lyric: Crashaw's Ritual, Liminal and Visual Wounds', in J.R. Roberts (ed.), *New Perspectives on the Seventeenth-Century English Religious Lyric*, Columbia and London: University of Missouri Press, pp. 237–67 (p. 250).
27. Pacheco, F., *Arte de la Pintura* (1638); cited in Carr (1964), p. 7.
28. See Andrews, L. (1995), *Story and Space in Renaissance Art: The Rebirth of Continuous Narrative*, Cambridge and New York: Cambridge University Press, pp. 2–3.
29. Pacheco, F. (1638); cited in Carr (1964), p. 51.
30. Camporesi, P. (1989), 'The Consecrated Host: A Wondrous Excess', in M. Feher with R. Naddaff and N. Tazi (eds), *Zone*, 3, *Fragments for a History of the Human Body*, New York: Zone, Part 1, pp. 220–37; esp. p. 222.
31. This notion emerged from exegesis of Revelation 6.9–11, especially the Old French gloss in the Apocalypse MSS, which is paraphrased in [Anon.], *Masterpieces of Spanish Painting* (1958), Norwich: Beaverbrook Newspapers Ltd in collaboration with Harry N. Abrams, Inc., p. 4.
32. For relevant aspects of Counter-Reformation eucharistic theology, see Cajetan, Cardinal T. de Vio, *De erroribus contingentibus in eucharistiae sacramento* (1525), esp. p. 44 and ch. 9, discussed extensively in Clark, S.J.F. (1967), *Eucharistic Sacrifice and the Reformation*, Oxford: Basil Blackwell, pp. 81, 86 and *passim*.
33. *The Canons and Decrees of the Council of Trent* (1848), pp. 235–6. Also see Hauser, A. (1968), *The Social History of Art*, 4 vols, London: Routledge and Kegan Paul, *Vol. 2, Renaissance, Mannerism and Baroque*, pp. 111–12.
34. Pacheco (1638); cited in Carr (1964), p. 51.
35. According to Karl Rahner, martyrdom is a sacrament: Rahner, K. (1961), *On The Theology of Death*, Quaestiones Disputatae, trans. C.H. Henkey, Freiburg: Herder; Edinburgh and London: Nelson, pp. 110–13; also see pp. 97, 109.
36. Stefaniak, R. (1992), 'Replicating Mysteries of the Passion: Rosso's *Dead Christ with Angels*', *Renaissance Quarterly*, **45**, 677–738, esp. 678–9, 693, 701–2.
37. Pacheco, F., 'Preface' to *Arte de la Pintura* (1638); cited in Hartley, C.G. (1904), *A Record of Spanish Painting*, London and Newcastle-on-Tyne: Walter Scott Publishing, p. 132.
38. Cajetan, Cardinal T. de Vio (1531), *De missae sacrificio et ritu adversus Lutheranos*, cap. 6; cited in Clark (1967), p. 90.
39. This interpretation of Pacheco's mannered treatment of the sagittation is indebted to Henri Zerner's psychological interpretation of the artistic idiom of mannerism; see Zerner, H. (1972), 'Observations on the Use of the Concept of Mannerism', in F.W. Robinson and S.G. Nichols Jr (eds), *The Meaning of Mannerism*, Hanover, NH: University Press of New England, pp. 105–21; esp. pp. 115–18. On the combination of styles to which Pacheco was indebted, see Muller, P.E. (1961), 'Francisco Pacheco as a Painter', *Marsyas*, **10**, 34–44.
40. The male figures illustrate pages 149 to 253 in the 1545 Latin edition of *De Dissectione*, which is the edition to which all references are hereafter made. Some

of these figures seem to be based upon the illustrations to Berengario da Carpi, J. (1521), *Carpi Commentaria cum amplissimus Additionibus super Anatomia Mundini una cum textu eiusdem in pristinum et verum nitorem redacto*, Bologna; but the 'Berengario series' is interrupted on page 168 and stops at page 170. For the sequence of events in the printing and publishing of Estienne's book, see Roberts, K.B. and Tomlinson, J.D.W. (1992), *The Fabric of the Body: European Traditions of Anatomical Illustration*, Oxford: Clarendon Press, p. 168.

41. Kellett, C.E. (1957), 'A Note on Rosso and the Illustrations to Charles Estienne's *De dissectione', Journal of the History of Medicine*, **12**, 325–36 (p. 327). These figures, except that on page 168, appear after the 'Berengario series' (see n. 40 above) and end at page 253.
42. The plates containing these figures are hereafter in the text referred to as the 'Sebastian plates'.
43. This critical term is proposed by Tony Bennett, and is here applied in accord with its explanation by Howard Felperin: Felperin, H. (1992), *The Uses of the Canon: Elizabethan Literature and Contemporary Theory*, Oxford: Clarendon Press, p. 97. All the plates in *De Dissectione*, together with three plates which had been discarded by Estienne, were republished by Jacques Kerver in Jollat (1557), *Les Figures et portraicts des parties du corps humain*, Paris; another edition was published in 1575.
44. Zupnick (1958), p. 81.
45. See, for example, Schultz, B. (1985), *Art and Anatomy in Renaissance Italy*, Ann Arbor, MI: UMI Research Press; Epping, UK: Bowker Publishing Company, pp. 80–81.
46. Heckscher, W.S. (1958), *Rembrandt's Anatomy of Dr. Nicolaas Tulp: An Iconological Study*, New York: New York University Press, pp. 87–8.
47. Sawday, J. (1995), *The Body Emblazoned: Dissection and the Human Body in Renaissance Culture*, London and New York: Routledge, p. 115.
48. Panofsky, E. (1963), 'Et in Arcadia Ego', in R. Klibansky and H.J. Paton (eds), *Philosophy and History*, New York: Harper and Row, pp. 223–54.
49. This plate occurs on f. 14 *verso* of (the unpaginated) Jollat (1575); see n. 43 above.
50. Paré (1678), p. 517.
51. Sawday (1995), pp. 113–14. Also see: Edgerton, S.Y., Jr (1985), *Pictures and Punishment: Art and Criminal Prosecution during the Florentine Renaissance*, Ithaca and London: Cornell University Press, pp. 157–64; Heckscher (1958), chs XIII and XV; Park, K. (1994), 'The Criminal and the Saintly Body: Autopsy and Dissection in Renaissance Italy', *Renaissance Quarterly*, **47**, 1–33.
52. See Jenner, M.S.R. (1999), 'Body, Image, Text in Early Modern Europe', *Social History of Medicine*, **12**, 143–54, which reviews Sawday (1995).
53. A case is made in Kellett (1957), for the basis of this series of illustrations of *De Dissectione* in a book of anatomical designs by Rosso Fiorentino, which the artist brought to Paris in 1530.
54. Edgerton (1985), pp. 134–5.
55. Several recent studies concur in this interpretation, and have augmented the reasons for adhering to it: Cunningham, A. (1997), *The Anatomical Renaissance: The Resurrection of the Anatomical Projects of the Ancients*, Aldershot: Scolar Press; French, R. (1999), *Dissection and Vivisection in the European Renaissance*, Aldershot, UK and Brookfield, VT: Ashgate.
56. Sawday (1995), p. 117.
57. Idem; also see pp. 113–14.

58. Choulant, L. (1962), *History and Bibliography of Anatomic Illustration*, trans. M. Frank, London: Hafner, p. 153. See Kellett (1957), pp. 334–5, on the alterations made to the woodblocks.
59. The citation is from French, R. (1994), *William Harvey's Natural Philosophy*, Cambridge: Cambridge University Press, p. 4. Also see: Cunningham (1997), p. 38; French (1999), pp. 215–29; Sawday (1995), pp. 117–18.
60. Ibid., p. 118. For a related discussion of the sacramentalism of dissection, see Edgerton (1985), p. 213.
61. Cited in Zupnick (1958), p. 129.
62. Henderson (1988), pp. 181–2.
63. Clapham (1604), f. A2 *r*.

CHAPTER SIX

Restoring all Things from the Curse: Millenarianism, Alchemy, Science and Politics in the Writings of Gerrard Winstanley

Andrew Bradstock

One of the ongoing debates in Winstanley scholarship concerns the extent to which he may be understood as a 'modern' – a thinker whose ideas belong to a later age. Winstanley's goal was a communist society in which the concept of private property would disappear and in which all would share equally the land and its fruits: with ownership, buying and selling outlawed, and true community restored, social and economic distinctions would be eliminated, and needs would be served by the common storehouse. In many ways Winstanley harked back, not just to the writings of Thomas More or the sermons of John Ball, but to the 'golden age' before the Fall, when all lived in a state of innocence and the earth was the 'common treasury', the Creator intended it to be.[1] Yet his ideas are also echoed by later communist and revolutionary thinkers, most obviously Marx, and more than one commentator has noted the relevance of Digger ideas to more recent periods. For many modern readers of Winstanley, as David Mulder observes, he was essentially 'a modern man trapped in the early-modern period'.[2]

Christopher Hill is perhaps the most notable of these: 'Winstanley had grasped a crucial point in modern political thinking', he notes in his introduction to the collection of Winstanley's writings he edited in 1973, 'that state power is related to the property system and to the body of ideas which supports that system.' Winstanley 'is modern too in wanting a revolution which would replace competition by concern for the community, in insisting that political freedom is impossible without economic equality, and that this means abolishing private property and wage labour.' Since Winstanley was writing before the Industrial Revolution, Hill suggests, some of his insights 'may be of interest to those in the Third World today who face the transition from an agrarian to an industrial society'.[3] The hint of what we would now call 'class consciousness' in Winstanley's writings also appears to suggest to Hill and others the possibility of reading Winstanley as a modern. This can be discerned in his concept of 'the kingly power', a collective term for the landlords, priests and lawyers who together maintained the iniquitous economic system which burdened the poor, and in his references (couched in biblical terminology) to the struggle between the 'young brother' ('Abel', 'Jacob') and the older ('Cain', 'Esau').[4] Yet, notwithstanding the appeal of such arguments, nor the fact that Digger ideals continue to inspire and inform land rights and environmental movements today, Winstanley has also to be seen as a person of his own time, one constrained by the thinking of his day about history, cosmology and religion. The 'Winstanley as modern' argument has to confront both his apparently mid-seventeenth-century understanding of the universe, and the biblical and mystical language in which he expressed so many of his ideas.

The extent to which attempts to 'modernise' Winstanley might be problematic in scientific terms has been brought out by David Mulder. In his highly original study of Winstanley as a person of his time, *The Alchemy of Revolution: Gerrard Winstanley's Occultism and Seventeenth-Century English Communism*, Mulder argues that, while there are clearly logical, rational and practical elements in Winstanley's programme to bring about a communist society in England, 'he was no more capable than any of his most conservative contemporaries of taking a straightforward approach to ... radical change'. In other words, while 'he envisaged a transformation of English society on a scale that we would call revolutionary, he could not conceptualise that transformation in modern, direct evolutionary terms'.[5] This was partly because Winstanley took seriously biblical prophecies concerning the second coming of Christ (as we shall discuss shortly), but also because, Mulder claims, his world-view, like that of his contemporaries, was informed by belief in magic, alchemy and hermeticism. Winstanley was in no sense a 'modern', since, the practical dimension to his programme notwithstanding, he was essentially 'striving to provide for the most basic needs of the very poorest sort of people through what we would call magic'.[6]

Mulder argues that, in much the same way that today we assume the existence and power of gravity, Winstanley took for granted the existence and influence of 'invisible spirits and essences'. His basic mode of thought was, in the precise sense of the word, 'occult'.[7] Mulder's Winstanley was very much influenced by hermeticism, a philosophy derived from the teaching of 'Hermes Trismegistus' and promulgated, in particular, by Jacob Boehme, which held that the alchemical practice of transmutating base metals into gold reflected the mystical operation of God's spirit in the universe. Like most of his contemporaries, therefore, Winstanley accepted a divinely ordered and controlled universe, and one, moreover, whose harmony and equilibrium could be affected by events in human history. An idea still lingering on from the Middle Ages, Mulder argues, was that any radical challenge to the status quo was also a radical challenge to the divine order of the universe: political revolution, for example, might precipitate 'an analogous reaction in the natural world, resulting in earthquakes, thunderstorms' and so on.[8] Where Winstanley differed from many in his day was in believing that the revolution he proposed would restore harmony to the universe rather than propel it into chaos: for him, the prevailing cosmology was a spur to social and economic transformation rather than a brake holding it back. But that he believed that his programme would have 'cosmological consequences' is not to be doubted.

It is at least arguable that the fulfilment of his project *depended* upon some kind of divine intervention. As commentators on Winstanley such as George Sabine and Paul Elmen have pointed out, the Diggers' project was unworkable in purely economic terms since it envisaged a large section of the population of England being fed from land which had hitherto not been arable.[9] Only because there was a theological and mystical dimension to Winstanley's programme, an identification of the communisation of the land with the lifting-up of the creation from bondage and restoration of all things from the curse, did he maintain his commitment to bring it about. Indeed, it is clear from his last pre-Digger tract, *The New Law of Righteousnes* (January 1649), that Winstanley understood lifting the creation out of its bondage to embrace not only the regaining of a true spirit of community among men and women, but the freeing of animals from exploitative use and the return of the land to its prelapsarian fertile state:

> When man-kinde shall be restored, and delivered from the curse, and all spirited with this one power, then other creatures shall be restored likewise, and freed from their burdens: as the Earth, from thorns, and briars, and barrennesse, the Air and winds from unseasonable storms and distempers; the Cattle from bitternesse and rage one against another ... And this is the work of Restoration.[10]

Or again,

> When this restoration breaks forth in righteous action, the curse then shall be removed from the Creation, Fire, Water, Earth and Air ... There shall be no barrennesse in the earth or cattle, for they shall bring forth fruit abundantly.[11]

'The earth is declared unfulfilled', as Nigel Smith summarises Winstanley's position, 'until it is a common treasury.'[12]

Yet Winstanley's apparent expectation that divine assistance, and perhaps also what Mulder calls 'magic', would help the Diggers' cause, should not obscure the fact that he also considered how *practically* he might realise his vision of establishing a communist society in the England of his day – and, indeed, beyond. Believing that his programme coincided with the divine purpose did not preclude the necessity to work to bring that purpose about –

in fact, quite the reverse, as his reflection at the beginning of *A Watch-Word to the City of London and the Armie* (August 1649) that 'action is the life of all, and if thou dost not act, thou dost nothing' makes palpably clear.[13] The whole tenor of Winstanley's writings and action ought to render it unnecessary to argue this point, though it has been occasionally suggested that he saw a merely symbolic role for his digging and anticipated the carrying through of the revolution by some external agent, usually God.[14] A recent extreme example of such a reading is that by John Rogers, which is premised on an assumption that Winstanley adhered to another philosophical movement of his day, the 'vitalist moment'. Arguing that Winstanley held 'all material substance' to be infused 'with the power of reason and self-motivation', Rogers posits, on the basis of his interpretation of Winstanley's reflections on Genesis 3.15 in *The Fire in the Bush*, that the Digger leader expected 'the murderous serpent of private property [to] be bruised not by direct human action or aggressive acts of political levelling but by a seed of "patience" and "love"'.[15] Rogers overlooks the rich terminology Winstanley uses for Christ. Furthermore, by failing to appreciate that the Digger's understanding that the Second Coming will take the form of Christ 'rising in sons and daughters' as they work the land together in love, he consequently invests the 'seed' with some mysterious power to carry through the revolution itself. As a result he is forced to construct a convoluted argument to explain the meaning of Winstanley's 'acting with Plow and Spade'. 'The hoped-for revolution', he concludes, 'would be the result, not of human effort, not even of divine providence, but of an organic process regulated internally by what Winstanley saw as an abstract principle of immanent divinity.'[16] If the price of making Winstanly a 'vitalist momenter' is to see his digging as no more than a 'gesture', and an ironic one at that, it is way too high!

For the Diggers were, of course, committed to direct and practical action – indeed, to transforming the very basis of land-ownership in England as 'communism' spread and the poor saw the rationality of the Diggers' system and left the employ of their masters. Therefore they looked to all workable methods of cultivation and husbandry to further their aims. New techniques aimed at making more of the land fertile – such as introducing fallow crops into regular field rotations and using the same land for both arable and pasture (so-called 'up-and-down' husbandry) – were becoming widespread in their day, and the Diggers not only adopted these themselves but would have recognised them as vital for the realisation of their vision. Common land was, after all, only common because its soil was poor and suitable mainly for grazing, whereas the planting of crops which could survive in dry and sandy soil; the use of those crops to maintain more animals and therefore produce more manure; and the pasturing of animals on land that would also be used, in rotation, for arable: all would gradually render land, hitherto infertile and unproductive, suitable for growing corn and other 'staple' crops. And that there was enough land with the potential to become fertile Winstanley had also worked out: 'Divide England into three parts', he would write in *The New Law of Righteousnes*, 'scarce one part is manured: so that here is land enough to maintain all her children ...'[17]

The evidence that Winstanley both knew of and adopted 'new' agricultural practices such as these is convincing. A manorial survey of Cobham conducted in 1598 suggests that some form of 'up-and-down' husbandry had been operating in the part of Surrey he knew for some 50 years before his arrival there, and had even been adopted to tackle the problem of reclaiming wasteland.[18] And from what we know of the Diggers' initial activities on St George's Hill, as recorded in the 'information' of Henry Sanders dated 16 April 1649 and Winstanley's own writings, it seems clear that, while they intended eventually to grow corn (and were to grow wheat and rye at Littleheath, Cobham), they began by sowing fallow crops like parsnips and carrots.[19] This was a practice also adopted by many of the other Digger colonies, the *Declaration* issued by the Diggers in Iver, Buckinghamshire, in May 1650, noting that 'they have begun mannuring the Commons in Kent, at W[e]llingborough, and Bosworth old in Northamptonshire & in Gloucestershire, & in Nottinghamshire, and they intend to sow roots til July, & then follow for winter corne ...'[20] Sanders' report and Winstanley's own account also suggest that, in the process of making the land fertile, the Diggers both manured and 'devonshired' (burned) it; and while there may be some substance to Mulder's assertion that Winstanley may have seen these processes more in terms of an application of the elements fire and water, and even the new agricultural methods he was adopting as 'a species of alchemy', it is nevertheless at least arguable that he also understood in practical terms how his vision might be accomplished.[21] Something of the seriousness with which he regarded such disciplines as husbandry and arboriculture is clear from his last tract, *The Law of Freedom in a Platform* (1651), on which he was working during the digging project: opportunities aplenty were to exist in the new society he hoped to see emerge for the pursuit of knowledge and understanding of these and other sciences.[22] As Carola Scott-Luckens points out in Chapter 7 in this volume, improvements in agricultural methods were proceeding apace in the seventeenth century, but within the framework of existing economic arrangements: what Winstanley envisaged was 'all the Commons and waste Ground in England, and in the whole World' being 'taken in and Manured by the People', who would thus be 'taking the Earth to be a Common Treasury, as it was first made for all'.[23]

At the end of the day, Mulder's line on Winstanley as a believer in alchemy, though strongly argued, is not as well-supported by Winstanley's writings as Mulder would wish: in fact, it is quite seriously challenged by some of Winstanley's own statements, such as his plea in *The Law of Freedom* that 'every one who speaks of any Herb, Plant, Art of Nature of Mankind … speak nothing by imagination, but what he hath found out by his own industry and observation in tryal'.[24] Mulder appears to operate on the premise that, since hermetic ideas were common currency in Winstanley's day,[25] he himself must have embraced them and they must have informed his words and actions. Mulder's claims seem largely to be based upon inference, and a 'hermetic' reading of Winstanley's tracts – whilst in theory valid – does little more than add emphasis to the argument that he cannot be read entirely as a materialist, or a 'modern' – an argument which, as I shall try to show, can be made much more convincingly by reference to the biblical resources upon which he draws. Yet Mulder does breathe new life into Winstanley scholarship by reminding us of the 'scientific milieu' in which the Digger leader was working and how, within that milieu, he could have unproblematically held an expectation that his digging would enable creation to break free from the limitations imposed on it by the Fall – even if that surprises or disappoints some who read him in a later era.

This reference to the Diggers' work releasing creation from its bondage – perhaps drawn from the eighth chapter of Paul's letter to the Romans[26] – is one of many clues in Winstanley's writings that he shared with his contemporaries in the radical Protestant tradition a conviction that biblical prophecies concerning the last days were being fulfilled in his own day. Winstanley is certainly clear that the restoration of true community which he sought was of a piece with the return of Christ, who would come, not in the form of a single person, but by rising up from the earth and indwelling and bringing together in community his sons and daughters.[27] This 'millenarian' strand in Winstanley ought, at least prima facie, to caution us against embracing him too readily as a 'modern'.

That millenarian ideas were rife in England during the 1640s and 1650s is none too surprising; among a people well-versed in the scriptures, and well-used to searching them for clues to the meaning and drift of contemporary events, it would have seemed almost axiomatic that the extraordinary happenings of those decades were of apocalyptic significance, that the Second Coming would occur during their lifetime. Milton spoke of Christ as the 'shortly-expected King'; Muggleton and Reeve announced themselves as the 'Two Last Witnesses' of Revelation, Chapter 11; Fifth Monarchists saw in the execution of Charles Stuart the fulfilment of the Danielic prophecy concerning the downfall of the world's four great historical dynasties; and any number of sermons were preached on the theme of Christ's return. Opinions varied as to the form this would take, but few people doubted it would soon occur.

There was a serious scientific and mathematical dimension to the business of dating the millennium. Considerable emphasis was placed, for example, on the number 1,260, which is mentioned explicitly in Revelation 11:3 and 12:6 and cryptically in references to 'a time, times and half a time' in other passages, this usually being taken to refer to a period of three and a half 360-day years. The twelfth-century seer Joachim of Fiore had divided history into three 'ages' and calculated that the Second Age would comprise 42 generations each of 30 years, a total of 1,260 years; and while he had been careful to emphasise the uncertainty of all calculations to do with the 'Last Things', many in the seventeenth century found his scheme a helpfully precise tool for discerning the signs of the times. Some believed, for example, that if (as Revelation 13 suggested) the Beast was allowed to exercise his power for 42 months, or 1,260 days or years, then by ascertaining when his 'reign' began one could work out when the next stage in the apocalyptic timetable might be likely to occur. And this was possible because, as Hill reminds us, the Beast was taken by most seventeenth-century protestants to be the Pope, whose translation from bishop of Rome to supreme pontiff was estimated to have occurred around either AD 390–96 or 400–406. The former calculation, as Hill puts it, 'pointed to the years 1650–56 for the destruction of Antichrist, the gathering of the Gentiles, the conversion of the Jews and their return to Palestine', whereas the latter would give the year 1666 especial importance in the prophetic timetable – a status enhanced, of course, by its similarity to the number of the Beast as detailed in Revelation 13:18. The year 1656 was also invested with some significance, corresponding as it did to the number of years which were understood to have elapsed between the Creation and the Flood.[28]

In his pre-Digger days Winstanley linked his conception of the millennium, the thousand-year reign of Christ over a perfect kingdom on earth foretold in Revelation chapter 20, to a 'dispensationalist' view of history, as the title of his first tract – *The Mysterie of God, Concerning the whole Creation, Mankinde, To be Made known to every man and woman, after seven Dispensations and Seasons of Time are passed over* … (1648) – makes clear.[29] According to this schema, as Winstanley explains in some detail in this work, biblical prophecies concerning the establishment of the kingdom of God on earth will one day be fulfilled, but 'in length of time, by degrees', God having been pleased to provide 'dispensations or discoveries of himselfe … which he will have the creature to passe through before he finish his work, to cast the Serpent, death and hell into the lake, and before he himself appeare …'[30] These dispensations are divisions of history, each marked at their beginning and end by an event of great religious significance, and each symbolic of the progressing or deepening relationship between God and God's creation. Winstanley, like all millenarians, is living in the penultimate (sixth) dispensation, during which the elect will be gathered together into one city and perfected. But he is also assured that the seventh is not far off, when the whole creation will be set free, and the previously unrepentant and wicked will enter the city of Zion, whither the elect have already gone. All humankind, as Winstanley constantly maintained, will ultimately be saved, with only the Serpent – whose defeat was prophesied during the first dispensation – in line for eternal punishment. Winstanley's knowledge of the ultimate fate of the Serpent is derived from Genesis 3:15, in which God promises to put enmity between it and Eve and her seed, and it is this metaphor of the Serpent, continually at war with humankind though awaiting ultimately its own final destruction, which gives Winstanley his clue to the proximity of the seventh dispensation, the millennium. He is wise enough not to suggest an exact date for its arrival – that remains the mystery of God – but he is clear that during the latter days of the penultimate dispensation, when the Serpent knows that its days are numbered, its rage will increase and manifest itself in 'reproaches, slanders, oppressions … and … temptations' directed at the saints.[31]

After *The Mystery of God* Winstanley's interest in dispensationalism appears to wane, though not his dependence upon the apocalyptic literature of the Bible to provide a key to understanding the signs of the times. The imagery of the Beast or Serpent, for example, representing those who oppose the work of the people of God, remains central to his thinking, and the biblical prophecies concerning its ultimate defeat continue to confirm to him that his hope is not in vain. Genesis 3:15 is cited repeatedly in his second tract, *The Breaking of the Day of God* (May 1648), and its theme appears in many of his subsequent tracts.[32] In *A Declaration from the Poor Oppressed People of England* (June 1649), the Beast appears together with its enigmatic number '666',[33] and this motif appears again in *A Humble Request* (April 1650) where Winstanley castigates Parson John Platt and the others who had violently forced the Diggers from the land as men who 'do so powerfully act the Image of the Beast'. Their days, however, are numbered too, for the work of digging, 'or the appearance of Christ in the earth … hath ripped up the bottom of their Religion', and 'when the Lamb turns into the Lion, they will remember what they have done, and mourne'.[34]

Despite an adherence to a dispensationalist view of history and a predisposition to interpret the present against the background of the apocalyptic drama set forth in scripture, Winstanley differed in two important respects from many of his contemporaries who adopted a millenarian position. First, although he never doubted he was living in the penultimate age or dispensation, he never offered a precise date or time by when he imagined the millennium would be installed; and that he avoids this sort of speculation in part arises from a second distinguishing feature of his millenarianism, namely his understanding of it in gradualist and immanentist, rather than metaphysical and triumphalist, terms. In the face of a widespread belief that Christ would suddenly and literally appear on Earth to reign in person – or, as Fifth Monarchists hoped, through his elect, the saints – Winstanley saw the Second Coming in terms of Christ 'rising up' as it were within men and women, in order gradually to restore 'right Reason' within their hearts and forge them together once more into a true egalitarian community. 'Christ in his first and second coming in flesh … is Justice and Judgement ruling in man', Winstanley wrote in *The New Law of Righteousnes*.[35] Winstanley is in fact explicit in his rejection of a literal, physical appearance of Christ, and from his earliest writings argued against portrayals of Christ as a man separate from the rest of humanity. 'Christ is not a single man at a distance from you; but … is the wisdome and power of the father, who spirits the whole creation, dwelling and ruling King of righteousnesse in your very flesh', he asserted in *The Saints Paradice* (1648?).[36] Or, as he had earlier put it in *The Breaking of the Day of God*, Christ should not be understood as separate from the saints, 'his body and spiritual house'.[37] Thus his equation of the second coming of Christ with the gradual transformation of humanity is consistent with his overall theological position. And it *is* an equation: for Winstanley, the work of digging was evidence that the 'return' of Christ had begun.

Winstanley also takes the wholly unorthodox step of equating Christ's second coming with his resurrection: there is no sharp contrast in his thinking, as Hill points out, between 'Christ rising in sons and daughters' and his 'descending from heaven at a much later date'.[38] To expect Christ to 'come in one single person', as Winstanley says in *The Saints Paradice*, is to 'mistake the resurrection of Christ'; 'you must see, feel and know from himself his own resurrection within you, if you expect life and peace by him …' And Christ is 'now rising and spreading himselfe in these his sons and daughters, and so rising from one to many persons, till he enlighten the whole creation (mankinde) in every branch of it …'[39] Winstanley's conviction was therefore that society would be changed gradually; the final age would arrive, not on the heels of a literal return of Christ in person, but as the result of a spiritual transformation of individual women and men. 'Christ's resurrection and second coming collapse into each person's sanctification', as David Dawson has put it.[40] And the effect of this, Winstanley claims on the basis of another original insight, will be true community or communism:

> when he hath spread himself abroad amongst his Sons and daughters, the members of his mystical body, then this community of love and righteousnesse, making all to use the blessings of the earth as a common Treasurie amongst them, shal break forth again in his glory, and fil the earth, and shal be no more supprest: And none shal say, this is mine, but every one shal preserve each other in love.[41]

Communal ownership and the resurrection of Christ were, for Winstanley, interchangeable concepts.

Despite Winstanley's immanentist understanding of the millennium, and apparent lack of concern about the exact date of its arrival, it is not to be doubted that a belief in its impending arrival informed his political programme. Some scholars have suggested that Winstanley's commitment to a millenarian position disappeared during the digging venture, and certainly by the time he came to pen his last and essentially utopian tract *The Law of Freedom in a Platform*, published in November 1651.[42] Undoubtedly the hostility which his digging colony experienced on St George's Hill, and Cromwell's suppression of the Levellers at Burford in May 1649, must have forced him to accept that he had underestimated both the 'institutional power of the Beast', and 'the hold of the serpent over the minds of [people]'. Perhaps these events also led him to conclude, as Hill suggests, that 'a period of education was needed before Christ arose in a sufficient number of sons and daughters to overthrow kingly power',[43] and even that the new society would have to be won with the assistance of those with real power. In any case, the appeals to the poor which had so characterised his early writings are replaced in his final work by a plea, indeed the dedication of the whole work, to the future Lord Protector: 'And now', Winstanley writes to Cromwell, 'I have set the candle at your door, for you have power in your hand … to Act for Common Freedome if you will; I have no power.'[44]

Yet if Winstanley had slowly lost the expectations he originally harboured of an imminent millennium by the time he came to write *The Law of Freedom*, his deep underlying hope that Christ would ultimately be seen rising in his sons and daughters and restoring again a true spirit of community remained. He still expresses there a hope that he is seeing the days of the resurrection to power of 'the spirit of universal Righteousness dwelling in Mankinde', and that in England,

> where this Commonwealths Government shall be first established, there shall be abundance of peace and plenty … There shall be no Tyrant Kings, Lords of Manors, Tything Priests, oppressing Lawyers, exacting Landlords, nor any such like pricking bryar in all this holy Mountain of the Lord God our Righteousness and Peace …[45]

In some respects this work may sound like an admission of defeat, a recognition that his hopes had been in vain, but it should also not be forgotten that it begins with the claim that 'the Spirit of the whole Creation … is about the Reformation of the World, and he will go forward in his work'.[46]

Winstanley's committed adherence to millenarianism, to a belief that his programme must be realised because of its central place in God's eschatological timetable, again suggests that it is most realistic to see him as a pre-modern figure. Though he was as far as he could be from seeing the struggle in which he was engaged as wholly spiritual, his conviction that the communist society he was working to achieve would materialise in a short time did appear to remove the imperative to take cognisance of the economic and political transformation occurring around him, to relate his programme to the actual historical conditions in which he was operating. Which is not to say, as we have argued earlier, that Winstanley did not embark on his programme without any reference to the prevailing circumstances, or with no grounds at all for supposing that his mission would meet with success: the call of God which he felt within him to the work of digging; the speed with which he was joined (in Surrey and farther afield) by others in whom Christ was also manifestly working; and the execution of Charles Stuart, which portended so clearly the imminent demise of the Beast in its many forms; these together provided him with enough 'empirical' evidence to sustain his belief that Christ was beginning to rise in his people, that the time for his programme was now. Winstanley also never saw Christ's coming as an isolated or disengaged event, an irruption into history without reference to social, political and economic conditions – for him it was intrinsically linked to the restoration of a communitarian form of social organisation. Yet, ultimately, the demands which he placed upon his eschatology, that Christ be brought back into history and *now*, did appear to lead him to believe that the reign of God, in the form of a truly communist society, could be achieved in his own day without reference to historical processes around him.

Perhaps the only way to rescue Winstanley as a 'modern' in the light of this would be to attempt to explain the biblical and mystical language in which he casts his discussion as simply a conventional way, for the time, of adding weight to his arguments, or a cloak to conceal his revolutionary intent – to draw a distinction, in other words, between the ideas that he espoused and the language in which he framed them. Hill appears to do this when he argues that Winstanley's ideas could be rewritten in the language of 'rational deism', and that had he lived 50 years later than he did he might have so expressed them.[47] In this Hill is not alone,[48] but the thesis is not unproblematic since, as we have shown, Winstanley seems firmly to have shared the milleniarian expectations of his day and to have allowed them materially to inform his politico-economic programme. And the premise that the language in which he expressed his political ideas might be incidental to the ideas themselves seems somewhat insecure, if not unhistorical, and conceivable only after an *a priori* commitment not to take him seriously as a person of his time.

As other essays in this volume have argued, it is far too simplistic to suggest that there was, in the early-modern period, a simple opposition of religious, and scientific and political thought. It is wrong to think that writers in the 1640s would automatically have excised religious terminology from their work had they lived into what we now call the Enlightenment. Such a view of Winstanley is sustainable only if the profound dialectical relationship that exists between his mystical theology and political vision is overlooked. In the final analysis it is not necessary to 'sift out' Winstanley's message from his medium, nor satisfy ourselves that he did not share the cosmology of his peers, in order to hear him today. There is still a freshness and power in his writings and action as he cuts through all that masks the root causes of human oppression and misery and shows us that authentic religion must orientate itself towards overcoming that oppression. For Winstanley saw the alienation of the poor, oppressed people of England as simultaneously political, economic and religious; he sought, in working to overcome this alienation, not to reject religion but to recast it in an immanent form: to show, in the end, the meek 'will inherit the earth'.

Notes

1. It is worth noting that Winstanley never acknowledges any source for his ideas: like all radical writers he claims to receive everything directly from God and his own reading of Scripture, and hence his only reference to a previous example of people living communally relates to an early church practice recorded in Acts 4:32: see Sabine, G.H. (ed.) (1965), *The Works of Gerrard Winstanley*, New York: Russell & Russell, p. 184. On Winstanley's attitude to books as a source of ideas, see ibid., p. 204.
2. Mulder, D. (1990), *The Alchemy of Revolution: Gerrard Winstanley's Occultism and Seventeenth-Century English Communism*, New York: Peter Lang, p. 4.
3. Hill, C. (ed.) (1973), *Winstanley: The Law of Freedom and Other Writings*, Harmondsworth: Penguin, pp. 9–10.
4. See, for example, Hill, C. (1993), *The English Bible and the Seventeenth-Century Revolution*, London: Penguin, p. 208; Turner, D. (1983), *Marxism and Christianity*, Oxford: Blackwell, p. 145; Mulder (1990), p. 74.
5. Mulder (1990), p. 47.
6. Ibid., p. 11.
7. Ibid., p. 4.
8. Ibid., pp. 48, 50.
9. Sabine (1965), p. 42; Elmen, P. (1954), 'The Theological Basis of Digger Communism', *Church History*, **23**, 216–17; Mulder (1990), pp. 191–2.
10. Sabine (1965), p.169.
11. Ibid., pp. 186, 199.
12. Smith, N. (2000), 'Gerrard Winstanley and the Literature of Revolution', in A. Bradstock (ed.), *Winstanley and the Diggers 1649–1999*, London, and Portland, OR: Frank Cass, p. 53. It is interesting to speculate whether there is any significance in Winstanley's choice of what appears to have been a particularly unpromising piece of common land to begin his project. Note, for example, the comment of one of the Surrey Diggers' contemporary detractors, William Blith, that 'if there be not thousands of places more capable of improvement than theirs, and that by many easier ways, and to far greater advantage, I will lay down the bucklers' (William Blith, 1652, *The English Improver Improved*, London); cited in Holstun, J. (1992), 'Rational Hunger: Gerrard Winstanley's *Hortus*

Inconclusus', in Holstun, J. (ed), *Pamphlet Wars: Prose in the English Revolution*, London, and Portland, OR: Frank Cass, p. 171; and Holstun, J. (2000), *Ehud's Dagger: Class Struggle in the English Revolution*, London and New York: Verso, p. 388. Note also Winstanley's own reference to the soil he was digging on 'George Hill' as 'very barren', in Sabine (1965), pp. 260, 333.

13. Sabine (1965), p. 315.
14. See, for example, Patrick, J.M. (1942), 'The Literature of the Diggers', *University of Toronto Quarterly*, **12**, 315; Hudson, W.S. (1946), 'Economic and Social Thought of Gerrard Winstanley: Was He a Seventeenth-Century Marxist?', *The Journal of Modern History*, **18**, 6, 11 and 21; Schenk, W. (1948), *The Concern for Social Justice in the Puritan Revolution*, London: Longman, Green & Co., p. 102; Elmen (1954), p. 213; Murphy, W.F. (1957), 'The Political Philosophy of Gerrard Winstanley', *Review of Politics*, **19**, 244, Knott, J.R. (1980), *The Sword of the Spirit: Puritan Responses to the Bible*, Chicago and London: University of Chicago Press, p. 86.
15. Rogers, J. (1996), *The Matter of Revolution: Science, Poetry and Politics in the Age of Milton*, Ithaca and London: Cornell University Press, p. 45; cf. p. 1.
16. Ibid., p. 48.
17. Sabine (1965), p. 200.
18. This survey is cited by Mulder (1990), pp. 198ff.
19. Firth, C.H. (ed.) (1992), *The Clarke Papers: vols I & II*, London: Royal Historical Society, pp. 210–11; Sabine (1965), pp. 257, 369, 393.
20. Hoption, A. (ed.) (1989), *Digger Tracts 1649–50*, London: Aporia Press, p. 33.
21. Firth (1992), pp. 210–11; Sabine (1965), p. 260; Mulder (1990), pp. 205–6. Winstanley makes frequent reference to the practice of manuring.
22. Sabine (1965), pp. 563, 577–79.
23. See Chapter 7 in this volume; also Holstun (2000), pp. 367–433; Sabine (1965), p. 260.
24. Ibid., p. 564. Occasionally Mulder appears to recognise his dilemma; see, for example, Mulder (1990), p. 53.
25. On this, see Thomas, K. (1973), *Religion and the Decline of Magic*, Harmondsworth: Penguin, pp. 267–70. Thomas notes that, despite losing their intellectual credibility by the middle of the seventeenth century, hermeticism and other magical ideas achieved considerable popularity during the Civil War and Interregnum.
26. Winstanley does appear to paraphrase this passage from Romans 8:22 in the *New Law of Righteousnes* (Sabine, 1965, p. 156).
27. In a sense Christ, for Winstanley, remains buried in the earth, which is why it is, for him, sacred. It has also a mystical quality: it is our 'true Mother … that brought us all forth' and that 'loves all her Children' (though she is hindered at present from 'giving all her Children suck' by the landlords' practice of enclosing the land and forcing the poor to starve; Sabine, 1965, p. 265). See also the discussion below of Winstanley's conception of Christ 'rising' in sons and daughters.
28. Hill, C. (1986), 'Till the conversion of the Jews', in C. Hill, *Collected Essays 2: Religion and Politics in 17th Century England*, Brighton: Harvester, pp. 270–72; Thomas (1973), p. 168.
29. Winstanley, G. (1648), *The Mysterie of God, Concerning the whole Creation, Mankinde, To be Made known to every man and woman, after seven Dispensations and Seasons of Time are passed over*, London: British Library, ref. 4377.a.51(1).
30. Ibid., p. 32.

31. Ibid., pp. 43–5.
32. See, for example, Sabine (1965), pp. 113, 117, 202, 253, 382, 453, 460, 486, 582.
33. Ibid., p. 270.
34. Ibid., pp. 436–47.
35. Ibid., pp. 204–5.
36. Winstanley, G. (1648?), *The Saints Paradice*, p. 116. This refers to the edition housed (on microfilm) in the Thomason Collection in the British Library, London, at reference E.2137. This edition is dated 1658, but since the work is included in *Several Pieces Gathered into one Volume* (1649) the date is either wrong or the edition not an original; see Sabine (1965), p. 91.
37. Winstanley, G. (1648), *The Breaking of the Day of God*, London: British Library, p.32 (ref. 4377.a.51(2)).
38. Hill, C. (1980), 'Debate: 'The Religion of Gerrard Winstanley; a Rejoinder', *Past and Present*, **89**, 148.
39. Winstanley, *The Saints Paradice*, pp. 82–4.
40. Dawson, D. (1990), 'Allegorical Intratextuality in Bunyan and Winstanley', *Journal of Religion*, **70**, 197.
41. Sabine (1965), p. 205.
42. Most notably George Juretic, who has argued that it is possible to speak of 'two Winstanleys', the early mystical millenarian and the mature, full-blown, secular communist; Juretic, G. (1975), 'Digger No Millenarian: The Revolutionizing of Gerrard Winstanley', *Journal of the History of Ideas*, **36**, 263–80.
43. Hill, C. (1978), *The Religion of Gerrard Winstanley*, Oxford: Past and Present Society, pp. 39–40; cf. Dow, F.D. (1985), *Radicalism in the English Revolution 1640–1660*, Oxford: Blackwell, p. 77; Sabine (1965), pp. 59–60.
44. Sabine (1965), p. 510.
45. Ibid., pp. 534–5.
46. Ibid., p. 502.
47. Hill (1978), p. 57. I offer a more detailed critique of the attitude of Hill, Juretic and others towards the religious language in Winstanley in Bradstock, A. (1991), 'Sowing in Hope: The Relevance of Theology to Gerrard Winstanley's Political Programme', *The Seventeenth Century*, **6**, 189–204; and in Bradstock, A. (1997), *Faith in the Revolution: The political theologies of Müntzer and Winstanley*, London: SPCK, pp. 82–107.
48. See, for example, Petegorsky, D. (1940), *Left-Wing Democracy in the English Civil War*, London: Gollancz, and Stroud: Alan Sutton, p. 206; George, C.H. (1975), 'Gerrard Winstanley: A Critical Retrospect', in R.C. Cole and M.E. Moody (eds), *The Dissenting Tradition: Essays for Leland H. Carlson*, Athens, OH: Ohio University Press, p. 214; and Aylmer, G.E. (1984), 'The Religion of Gerrard Winstanley', in J.F. McGregor and B. Reay (eds), *Radical Religion in the English Revolution*, Oxford: Oxford University Press, p. 95.

CHAPTER SEVEN

Providence, Earth's 'Treasury' and the Common Weal: Baconianism and Metaphysics in Millenarian Utopian Texts 1641–55

Carola Scott-Luckens

The decades immediately prior to the present new millennium have seen some valuable reassessments of earlier assumptions about the significance of millenarian and eschatological expectations in the development of early scientific thought and methodologies. Of particular interest are the 1640s and 1650s in witnessing the final attempts by intellectuals under the influence of the new thinking, to incorporate its ideas of exploiting empirical knowledge of the created world within a larger and comprehensive programme of religious, or metaphysical, philosophy. Despite a prevailing acceptance of the theological dictum that, consequent upon the Fall, all human knowledge was necessarily tainted and partial, there were many at this time who for various reasons were anticipating the coming of a new historical age of divine reconciliation and redemption, when human understanding and management of the natural world would be enabled to achieve its final and full potential within a regained state of edenic harmony.

Such thinking provides a useful context for viewing such ostensibly dissimilar enterprises as the Digger community of 1649–50 and the government-sponsored initiatives for social and economic improvement led by the intelligencer Samuel Hartlib. Utopian and experimental reformers of the mid-century broadly shared a belief that to investigate and utilise the natural world was to glorify its Creator, and potentially to recover that easy access to the fruits of the earth forfeited by a fallen humanity. Studies undertaken during the 1970s of philosophical and radical writings of the revolutionary period reveal a close association between Baconian precepts, Paracelsian concepts and Comenian *pansophia*. This philosophical linkage suggests that Protestant millenarianism and eschatology, a widely-held belief that the thousand-year reign of Christ prefigured in Revelation 20:1–5 was in some way historically imminent, provided the impetus both for ecclesiastical and political changes and for a variety of programmes of universal reformation of this period. It supports the premise advanced by Charles Webster, and more recently endorsed by others, that the reformist works produced by members of the Hartlib circle and framed on Comenian principles, represent an important link between Francis Bacon and the Royal Society.[1] Such interpretations run counter to the more traditional tendency, prevalent among Hartlib's post-Restoration critics, to dismiss initiatives such as those of the Hartlib circle and Robert Boyle's 'Invisible College' as too ambitiously metaphysical and diffuse in their remit to supply the prototypical model for the organisation later made famous by John Wilkins and Thomas Sprat. The continuance of this prejudice among certain twentieth-century historians of the Royal Society will be more fully discussed below.

To return briefly to Winstanley, the radical movement of democratic and community reform which he promoted centred upon the popularly-based ownership and usage of land as the optimum means of attaining greater national prosperity and a new and more just social order. But I believe that Winstanley's vision (dealt with more extensively in Chapter 6 in this volume) can be located within a wider spectrum of Baconian, pansophic and millenarian ideas which also informed the reformist works of Samuel Hartlib and his associates. In positioning Winstanley's view of land reform alongside Hartlibean utopian literature my object will be to re-examine ways in which the development of new intellectual philosophies were influenced by millenarianism, utopianism, and religious reformist ideals initiatives, making it possible for contemporaries to view the more effective use of natural resources by agricultural and technological innovations as part of a wider and vastly ambitious reorganisation of knowledge. With this premise in mind, Winstanley's project and the reform programmes of the Hartlib circle do exhibit certain common features. Both initiatives relied heavily on gaining the support of the authorities as well as the wider public. Also each perceived the natural world as a 'common treasury', the riches of which needed to be drawn upon for the wider good. To Winstanley this concept symbolised the material and spiritual restoration of an edenic (and post-'Norman') unity, in which improved agrarian techniques were to be implemented through a communitarian system of land ownership and usage. As will be shown below, Hartlibean enterprises also envisaged a restoration of prelapsarian harmony between nature and humankind, which bears closely upon Bacon's own perspectives of a coming new age of science:

> For creation was not by the curse made altogether and for ever a rebel, but in virtue of that charter 'In the sweat of thy face shalt thou eat bread', it is now by various labours ... in some measure subdued to the supplying of man with bread [and] to the uses of human life.[2]

Winstanley and changing methods of husbandry

Like other radical dissenting groups of the 1640s, the Diggers were convinced of their power to bring about by example the new world of Christ's second coming. More specifically their goal was to right the grievances of those dispossessed of the means to make a living from the land. *The True Levellers Standard Advanced* (1649) describes Gerrard Winstanley's early radical dream of a land-based community in which he and other Diggers could cultivate waste and common land. Like the utopian lands envisaged by More, Bacon and Hartlib's colleague Gabriel Plattes (see below), this was to be a patriarchal society in which authority was conferred upon male heads of families. The Diggers sought to awaken others to see Creation as having been 'made a common Treasury for all', a source of sufficiency for the needs of all by means of shared labour rather than the present site of competitive struggle: 'through mans unrighteous actions one over another ... a place wherein one torments another'.[3] Winstanley argued that it will not be long before 'the Earth becomes a Common Treasury again, as it must, for all the Prophecies of Scriptures and Reason are Circled here in this Community'.[4] Agricultural enclosure was especially condemned as 'that cursed Bondage' that deprived communities of their primal bond with the land: 'Therefore do not thou hinder the Mother Earth, from giving all her Children suck, by thy Inclosing it into particular hands ...'.[5]

Winstanley was not the first to accuse greedy landowners of depriving tenant farmers and smallholders of their livelihoods. Over a century earlier Thomas More's *Utopia* had denounced the 'great number of gentlemen' who lived on the industry of their tenants 'whom they poll and shave to the quick by raising their rents' until these dependants were reduced to starvation and vagrancy.[6] But the development to which Winstanley refers was part of a major, if gradual, shift in land management throughout the early-modern period, from communality to more profit-based cultivation techniques and centralised markets; changes which he saw as causing widespread poverty, civil strife and social injustice.[7] Evidence of these pressures is seen in husbandry manuals from the late sixteenth century, in which new, emergent discourses of agrarian improvement encouraging the aspirations of 'thrift-coveting' farmers and entrepreneurial landlords to maximise their profits involve changing from tied to waged labour. Yet the time-honoured model of local communities bound by networks of social duties and responsibilities continued to have cultural resonance, as exemplified in Ben Jonson's 'To Penshurst' (1616). The resultant dislocation between ideal and practice can be seen in cases such as that of Henry Percy, Ninth Earl of Northumberland. Advising his son in 1609 concerning estate management, Percy urges his heir to maintain the 'calling' of the traditional landed gentleman:

> it is not my meaning to make you a slave to your wealth, or a whole acting instrument to your profit; for that were too base, too much tasting of the clown, and loss of time from more worthy matters that your calling and place should move you to. For you to sit at the helm of your estate to direct well with expedition and ease, will be a means of upholding your honour with a good report.[8]

Nonetheless, Percy's own landlordship had brought about a dramatic recovery from a previous period of profligacy and decline. Under his management the net annual income from his estates more than doubled, increasing from £3,000 to £6,650 by the turn of the century.

At this time also increasing numbers of cheap pamphlets were making use of proverbs and jingles to impart information to the semi-literate small farmer. Above all, a newly-fashionable rhetorical emphasis in these manuals upon words such as 'order' and 'thrift' signalled a rising concern with property rights and the accumulation of personal wealth. In a period of improved land technologies and increasing regional specialisation, the profit motive was fast becoming respectable, while those without the means to own land were excluded, unless working as waged labour, for whom income levels were set by landowners.[9] Generally speaking then, these husbandry manuals seem to suggest that even in late Tudor England the concept of nature as a communal storehouse was already being superseded by the pressure of market forces and opportunities.

Nonetheless, the ideas drawn upon by Winstanley, of the natural world's 'treasury' as a specifically *public* resource, to be unlocked and utilised through human ingenuity and industry, were appearing in manuals on land usage almost a decade earlier than the Digger enterprise. *The Law of Freedom in a Platform* (1651), Winstanley's final blueprint for an ideal commonwealth based on a communal system of law, knowledge and practice, is but one of a range of contemporary texts which (while holding to separate and individual utopian models) conceive their theories of social organisation in the spirit of a fundamentally new philosophy. These include the pamphlet *Macaria* published by Samuel Hartlib (1641), Thomas Hobbes' *Leviathan* (1651), and James Harrington's *Commonwealth of Oceana* (1656).[10]

Baconianism and *Pansophia*: the emergence of the 'practical' utopia

Samuel Hartlib, the Anglo-Polish son of a merchant family in the Baltic town of Elbing in Prussia, first came to England to complete his studies at Cambridge in 1625–26. Here, among wealthy supporters of the Cambridge Platonists, he found patrons for his own embryonic intellectual network that later became the Hartlib circle. Permanently resident in London from 1628,

Hartlib carried on an increasingly wide foreign correspondence to promote utopian programmes of learning, together with religious, economic and social reforms.[11] John Milton saluted him in 1644 as 'a person sent hither by some good providence from a farre country to be the occasion and the incitement of great good to this Land'.[12] But it was Hartlib's role in fostering the collection and exchange of innovative ideas through the countries of Ireland, Europe and Transylvania to the colonies of New England, that caused John Winthrop Jr to title him 'the Great Intelligencer of Europe'.[13] Until recent years the scope and significance of Hartlib's written enterprises has been underappreciated due to the technical difficulties of gaining access to all his collected writings.[14] He is now recognised as foremost among a generation of European-born intellectuals who in the devastation of the Thirty Years War sought refuge in England, from which to respond to the Baconian proposals for a radical restructuring in the gathering and classifying of knowledge.

Bacon was anti-Platonist in perceiving knowledge as separate from being, and divided into different levels and types of discourse.[15] Following his early major work *The Advancement of Learning* (1605), the new inductive methodologies outlined in *Novum Organum* (1620) were intended to form the basis of his uncompleted *Instauratio Magna*, a hugely ambitious collective enterprise to coordinate and classify empirical knowledge as part of a worldwide intellectual regeneration.[16] The movement from interior to exterior realities embodied in *Novum Organum* is borne out in the discursive development of Bacon's *New Atlantis* (1627), in which constant journeyings reflect the actual explorations by which new lands were currently being revealed and exploited. It offers a demonstrably colonialising narrative, in which evidence gathered from a 'purer' society on the far reaches of the known world is brought back to revitalise the old, corrupt state. In the voyagers' fictional wanderings theological assumptions are confronted by the claims of science, with the latter emerging as the potential source of an achievable completeness.[17] Further philosophical linkages can be found between Bacon's experimental histories and logical methodologies and such works as Jacobus Acontius's *De Methodo* (1558) and Joachim Jungius's *Logica Hamburgensis* (1635), as well as with the later pansophic project of the Czech Protestant exile Jan Amos Komensky (Comenius).[18] Comenius' philosophy viewed human beings as formed in God's image, and therefore able by divine will to strive to achieve perfection in society and the world. He held that an inextricable bond existed between public and private spheres of life, and that the former was attainable only through the universalised betterment of each individual in society. But Comenius saw this goal as achievable only by an ambitious, prolonged and laborious enterprise: a comprehensive programme of educational reform.

How such an ideal might evolve is a core theme of the scientific utopia, of which genre Bacon's *New Atlantis* is an early example. Here Salomon's house (fictionally distanced from everyday reality on an island amid stormy seas) symbolises Bacon's views of the role of science, in which the state adopts an active role in funding coordinated scientific research by an elite brotherhood of scholars. By the 1640s, in a nation divided by religious disputes, political power struggles, and economic decline, Bacon's optimistic world-view was inspiring a growing community of pioneering philosophers, which included the international network of experimentalists built up by Samuel Hartlib, who sought to bring pressure upon state and entrepreneurial interests to support programmes of scientifically-based social and economic reform.

The adoption of Baconian principles within a context of active government involvement and support forms the basis of an anonymous but influential millenarian utopian tract published by Hartlib in 1641 (to coincide with the opening of the second session of the Long Parliament), *A description of the famous kingdome of Macaria*.[19] The author is thought to have been Gabriel Plattes, a practical experimenter and writer of several works on innovations in husbandry and mining, including an agricultural manual entitled *A Discovery of Infinite Treasure, Hidden since the Worlds Beginning* (1639). *Macaria*'s author also revealed his intention to publish a future great work entitled *The Treasure House of Nature Unlocked*, with parliamentary backing, to ensure that the rewards of scientific enterprise 'be made common to all'.[20]

In a fictional framework similar to that of More's *Utopia*, *Macaria* comprises a short dialogue between a scholar and a traveller recently returned to England. From discussion of the latter's travels, the description emerges of an idealised but essentially practical utopia: 'In a Kingdome called *Macaria* the King and Govenours doe live in great honour and riches, and the people doe live in great plenty, prosperitie, health, peace, and happinesse, and have not halfe so much trouble as they have in these European countries.'[21] In this new model society, inquiries into agriculture, technology and medicine are funded and coordinated by the state, ensuring that the fruits of these new discoveries are most effectively employed for the wider public good. Chief of the five councils in Macaria supervising research and development is the Council of Husbandry, by whose efforts 'the whole Kingdome is become like to a fruitfull Garden'.[22] Notably, there is an attempt to link property rights with the obligation to make a return in terms of food grown for all. The traveller, as authorial persona, resolves to publicise these new techniques in England: 'I will propound a book of Husbandry to the high Court of Parliament, whereby the Kingdome may maintaine double the number of people, which it doth now, and in more plenty and prosperity, than they now enjoy.'[23]

In response to the needs of the present time the pamphlet's account is designed to reach a broad popular audience through the medium of an emergent print trade: 'the Art of Printing will so spread knowledge, that the common people, knowing their own rights and liberties will not be governed by way of oppression; and so, little by little, all Kingdomes will be like to *Macaria*'.[24] What is outlined here seems little less than an ideal of peaceful economic and social revolution. Within the narrative dialogue the scholar is convinced of the superiority of the *Macaria* account over similar treatises of the past, with a practicality and conciseness uniquely suited to contemporary needs:

> I have read over Sir *Thomas Mores Utopia*, and my Lord *Bacons New Atlantis*, which hee called so in imitation of *Plato's* old one, but none of them giveth mee satisfaction, how the Kingdome of England may be happy, so much as this discourse, which is briefe and pithy, and easie to be effected, if all men be willing.[25]

Macaria was followed by a wave of works on reform in the 1640s and 1650s. By 1649 bad harvests and civil wars had brought England near to economic and social collapse, leading the new Commonwealth government to institute a pooling of state-sponsored intellectual and technological initiatives. In 1650, an economic programme was launched designed to stimulate domestic trade and manufacture, regulate imports and exports, and reduce unemployment. By the request of the Council of Trade, Hartlib circulated his own manuscript of propositions 'for advancing of agriculture' in May 1649, subscribed to by Robert Boyle and Sir Cheney Culpeper; and between 1650 and 1655 he was involved in the editing or publishing of over a dozen books on agricultural subjects.[26]

The Reformed Common-wealth of Bees (1655), produced at the high point of Hartlib's husbandry research, is among the most sophisticated of his own agricultural works. Beekeeping guides were a genre in which 'mini-utopian' aims could be advanced within the essentially practical remit of the husbandry manual. Following ancient models of husbandry from works by Pliny, Varro and Virgil, they underlined the importance in rural life of beekeeping, which in Bacon's *New Atlantis* featured alongside silkworm farming as profitable forms of insect-based enterprise. Apiculture was one of many reform schemes pursued by Hartlib in his programme to bring about national self-sufficiency; a similar work on silkworm culture is also appended to *The Reformed Common-wealth*.[27]

Given Hartlib's Comenian sympathies, one might expect the utopian programme of practical beekeeping of *The Reformed Common-wealth* to be presented as conformable with a moral (if not overtly political) framework. Traditionally the hive afforded a potential model of human society, by which the nature and function of communities might be better understood. The fourth book of Virgil's *Georgics* was particularly influential in acquainting classically-educated readers with the value of the beehive as a metaphor for society's harmonious industry and its exploitation. Both classical and Christian thought held that a knowledge of the divine might be acquired by means of a careful study of the natural world, and bees were reputed to be of impeccable ethical character: sound economists, good husbandmen, clean, chaste and industrious, and excelling in geometric and architectural skills. The social structure of the hive was a model of rational organisation, a true commonwealth in which property and labour were held strictly in common. As one William Mewe, a minister and experimentalist of Gloucestershire, wrote in a letter published by Hartlib: '[God] that sends us to the Ant, gives us leave to observe the same and better qualities in the Bee ... I observed many rarities in ther work and government.'[28] In *Commonwealth of Bees* Hartlib was also promoting an invention by Mewe: a new form of hive with a glass panel through which bee activities might be studied for moral and political lessons: 'The Invention is a fancie that suits with the nature of that Creature, [in that] they are much taken with their Grandeiur, and double their tasks with delight.'[29] Observers were fascinated by the way in which, amid the fall of human governments in the 1640s, the bee remained true to its own internal system of authority. But this innovation had also a more pragmatic purpose. One transparent hive yielded fourteen quarts of honey, doubling conventional output, so that its inhabitants 'quickly paid the charges with their profit, and doubled it with pleasure'. Moreover, Mewe reported that he was able to 'take strict account of their work and thereby guess how the rest prosper'.[30]

Hartlib's own beekeeping projects of the 1650s had as a practical goal the widespread increase of the nation's honey production. Already valuable for domestic consumption, honey was potentially a lucrative source of profits abroad, and Hartlib saw the expansion of English and colonial honey production as a vital contribution to solving England's economic crisis. While few of his proposals were taken up in any systematic way, a number of fellow experimentalists working in this area nationally and overseas welcomed Hartlib's collation of information on this and other topics as highly constructive: 'Sir, Being much indebted to you for the gift of your Legacie, and other choice pieces ...'; 'I commit you, in these your pious Endeavours of the Publique good, to his protection, who will undoubtedly at present blesse you in this your Enterprise ...'; and more portentously perhaps, 'your most Ingenious and Publique Spirit makes me love and honour you; onely I fear your sweetnesse may be abus'd by some undertakers that are apt to promise much upon the score of hopes and fancies ...'.[31]

Hartlibean reform and the Comenian dimension

Hartlib's concepts of beekeeping were arguably central to his wider vision of national agricultural reforms in the 1650s, for by introducing a degree of institutionalism and cooperation to researches on apiculture, *The Reformed Common-wealth* represents a step in the development of the modern scientific method. Its apparent exclusion of the direct analogic dimension prominent in some of the author's earlier works parallels Jon Jonston's major work *Historia Naturalis* (1650), seen as a major indicator of the downfall of natural histories based upon resemblance, portrayed by Foucault as part of the breakdown of the Renaissance *episteme* based on the 'order of similitude'.[32] This unifying and transcendent discourse, with its language and hermeneutics operating by a straightforward linkage between word and thing, signifier and signified, gradually gave way to models of relativised meaning within the systemic whole of a language-pattern or record of knowledge. The later rise of rationalism becomes part of this shift, as does the development of an emergent discourse in which allegory is more rigidly divided from observation and classification.

This perceived historical shift away from allegory, however, did not prevent the continuance of the hive-construct as a potent social metaphor. In 1705, Bernard Mandeville's 'The Grumbling Hive, or Knaves turn'd Honest' advanced the controversial assertion that a 'virtuous' society which sought merely its own bare subsistence would collapse from lack of trade and industry, whereas acquisitiveness and love of luxury served to fuel the economy by means of employment and wealth-creation:

> Thus every Part was full of Vice
> Yet the whole Mass a Paradise:
> Flatter'd in Peace, and fear'd in Wars,
> They were th'Esteem of Foreigners,
> And lavish of their Wealth and Lives,
> The Balance of all other Hives.
> …
> The worst of all the Multitude
> Did something for the Common Good.[33]

Timothy Raylor fastens on Mandeville's use of the term 'fable' to invalidate the analogic significance of his illustration, but the political and moral storm the work stirred up in its time belies this easy dismissal.[34] Thomson's *The Seasons* (1720s [1746]) offers a further dramatic parallel: of the 'still-heaving hive', formerly 'a proud city, populous and rich', its 'happy people in their waxen cells /… tending public cares and planning schemes / Of temperance for Winter', as a metropolis now overthrown by the 'dread earthquake' of human autumnal plunder in clouds of evening sulphur.[35]

That Hartlib's beekeeping programme, like his wider visions of structural reform in English education and scholarly and experimental research, remained unfulfilled, was probably due less to the decline of allegory than to the sheer comprehensiveness of the pansophic project. Critics seized upon its ambitious design for encyclopaedic learning as not making sufficient allowance for the contingent or divergent, as striving 'too much for compendiousnes and brevity, wheras some things must of necessity bee handled at large'. Comenius, it was said, 'supposes a world of th[in]gs and takes them for granted ... His Synthetical Method which hee follows spoils all. If there bee not a perfect gradation so that but one bee missing all is spoiled.'[36] Similarly, our own time has showed a tendency to evaluate early forms of scientific inquiry specifically in terms of their capacity for empirical detachment. On this basis a number of so-called 'ethical philosophers' whose investigations were carried on during the 1640s and 1650s have been dismissed by historians such as Hugh Trevor-Roper and Margery Purver from consideration as forerunners of the Royal Society.[37] Purver actually labels Hartlib as the 'originator' of Robert Boyle's Invisible College, despite evidence that these two scholarly communities were separate, if collaborative.[38] Hartlib's Comenian sympathies count heavily against him here, although a certain ambivalence can be detected. For instance, another modern critic gives a detailed analysis of the political and intellectual reformist influence of the pansophic alliance between Hartlib, Dury and Comenius, but insists on maintaining a distinction between their 'vulgar Baconianism' and the 'true Baconianism from which the Royal Society drew its philosophy'. In another broadly disparaging portrayal the same author does admit that 'in [Hartlib's] hopes for the foundation of a great philosophical college, and for the advancement of technology, he was at one with aspects of the Baconian tradition'.[39]

How are we to make sense of these anomalies? Foucauldian theory views Francis Bacon as leading the empirical offensive against similitude as truth source, in favour of the proto-scientific method based upon systematisation.[40] The reduction of all measurement to serial patternings beginning with the simplest forms, enables apparently random differences to be understood by their degree of complexity; but this model of comparison grounds the body of our acquired knowledge in a fundamental relativity and contingency. Concerned only with the method by which things can be known, such a system is hostile to any notion of an essence or core 'being' of things. It is surely this aspect of the developing scientific method which groups such as the followers of Comenius and Paracelsus and the Cambridge Platonists sought to address.

The extent, therefore, to which attempts by members of the Hartlib circle to construct practical applications of the new methodology modelled on

Comenian principles played an innovative part in the evolution of theories of meaning and knowledge remains an open question. Hartlib and his colleagues conceived an embodiment of reality in which different spheres of human society and nature were ruled by complex integrative systems.[41] However anachronistic to modern sensibilities, their genius for systematisation of knowledge and reform initiatives was very much in keeping with the types of wider exploration and applications of Baconian principles which were being undertaken in the 1640s and 1650s. Arguably the universalising reformist philosophy that informed Hartlib's agricultural and economic programmes, no less than his theories on language teaching or state-funded, international intellectual colleges, represents a development of scientific perspectives which had considerable import and influence in its time. Such a view is supported by one exponent of Hartlib's systemic philosophy:

> What happens in the works of Comenius and the scholars of the Hartlib circle is not a nostalgic return to 'the order of similitude', but *an attempt to construct a different and more functional semiotic order* under the pressure of rapidly developing empirical and experimental science, represented in England by the works of Francis Bacon.[42]

Hartlib saw language as the referential source for an intricate network of interlinked knowledge-patterns encompassing spiritual and temporal spheres. Given that the seventeenth-century mindset was not subject to the watertight compartmentalisation preferred by modern thought, such complex metaphysical systems might reasonably have provided a stimulus to scientific activity, giving rise to the kind of rigorous technological and analytical inquiry seen in a work such as *Macaria*:

> Transmutation, universal medicines or perpetual motion, although ultimately bypassed by advanced science, were not inherently unscientific hypotheses. Neither was an animistic worldview less conducive to scientific progress than the mechanistic philosophy ... The scientific community comprised men with fluctuating and wide-ranging interest; their minds moved rapidly and naturally from religious doctrine to scientific theory and technical innovation. It is quite arbitrary to assume *a priori* that any one of these facets of thought was irrelevant to the individual's scientific accomplishments.[43]

Webster's suggestion here, of a closer linkage in seventeenth-century thought between metaphysical and materialist realms, is borne out in the case of a major thinker of the time, Thomas Hobbes.

Protestant eschatology and the role of the state: bridging the metaphysical divide

Hobbes's *Leviathan* is recognised today as a materialist and universalising analysis of power in society, as set out in Books I and II. But Books III and IV place this rational construct of an ahistorical commonwealth within the larger frame of a specifically historicised world of faith, as articulated within Christian commonwealth. Revealed religion, according to Hobbes, is transmitted by divine prophecy, and can be understood only in terms of a linear and eschatological progression of divine acts said to have been performed by God in the past or promised by him in future.[44] In marked contrast to the universal moral statutes of 'Reason; the Dictates whereof are Laws, not *made* but *Eternall*', those decrees which are instituted 'by God himselfe ... are of the nature of written Law', that is, historically and particularly revealed in order that 'no man can excuse himself, by saying, he knew not they were [God's]'.[45] These revelations of divine authority might potentially come into conflict with civil philosophy and natural reason. Nonetheless, since faith gains its authority from both the content and transmission of revelation, this content must itself fall under the jurisdiction of the civil magistrate.[46]

In *Leviathan*, prophecy and eschatology are revealed as not merely a system of dogmas for believers, but a central part of the shared conceptual consciousness of English and continental Christianity. The work's first appearance in print coincided with the highest point of millenarian and eschatological expectations (of which the works of Winstanley and the philosophic aspirations of Hartlib, Dury and Comenius were also part) so it is not surprising to find these developments having a strong influence on Hobbes's own work. Although rarely seen as a religious work, *Leviathan* offers a significant endorsement of Protestant millenarian beliefs as located uncompromisingly in a materialist present and future. Peter's words of the New Testament serve to underline Hobbes's own rejection of a heaven and earth that exist outside of time:

> he saith, that the Heavens and the Earth that are now, are reserved unto fire against the day of Judgment, and perdition of ungodly men ... wherein the Heavens shall be on fire ... and the Elements shall melt with fervent heat. Neverthelesse, we according to the promise look for new Heavens, and a new Earth, wherein dwelleth righteousnesse.[47]

Like his view of the soul, Hobbes's conception of earth and heaven is fundamentally materialist, located in the world of a present and infinitely future time. As such, it can be seen as representing a philosophical link with the millenarian perspectives of Winstanley and other radical sectarians.[48] I would argue that it also applies to the utopian goals of Hartlib and Comenius.

Protestant eschatology's location of human salvation within linear history, and with its outward organisation subject to temporal authority, is in direct opposition to the traditional Catholic/Augustinian stress on more allegorical representations of the soul's redemption and ascent to heaven. By his use of eschatological rhetoric and doctrines, Hobbes sought, like Winstanley, to undermine the traditional power of church and clergy as exempt from civil jurisdiction. His view of individual salvation, and all aspects of spiritual revelation, as inextricably bound up with the material conditions of human temporal existence, and necessarily involving the authority of the state, shows an innovative attempt to grapple with the issues upon which Samuel Hartlib and his associates also sought to construct a new and better existence for the nation and its inhabitants.

It is the historical power of the pansophic vision, as much as its tangible links with modern scientific thinking, that must concern those wishing to gain a clearer understanding of these developments. The perceptions of philosophers and experimentalists of the mid-seventeenth century were conditioned by a world in which forms of intellectual and practical knowledge were becoming disturbingly fragmented and specialised. Hartlib's adoption of Comenius' logically-structured conception of the totality of human knowledge can be understood as a response to this dilemma. Significant in *pansophia* is its overall emphasis upon unity: the unification of the Protestant churches, the interconnection of the physical sciences, and the interpenetration of the material and spiritual worlds. Hartlib became a principal supporter of Comenius' theories as a means of strengthening the Reformation in England, by their translation of revolutionary theology into a providentially-sanctioned intellectual and moral renaissance. It was Hartlib's view that the formation of new systems of knowledge and language-teaching, together with the fostering of an international scholarly fraternity whose knowledge could be popularly disseminated through programmes of state patronage, would lead to an overall increase in national prosperity. Finally, in association with other aspects of his thinking on the materialist basis of culture, Hartlib's concept of the circularity of wealth both foreshadows and challenges twentieth-century economic theory. An enlightened government's ability to take part in the open and easy exchange of data and ideas, on national and international levels, was viewed by him as a form of 'cultural capital' just as real as the physical flow of money and goods. By its ability to bring about a change in people's thinking, he saw this freedom of information as vital to creating the incentive for social change and paving the way for future projects of economic and spiritual transformation.[49]

Notes

1. See Webster, C. (1975), *The Great Instauration: Science, Medicine and Reform 1626–1660*, London: Duckworth; also Webster, C. (1974), 'The Authorship and

Significance of *Macaria*', in C. Webster (ed.), *The Intellectual Revolution of the Seventeenth Century*, London: Routledge & Kegan Paul, pp. 369–85. Webster's view is echoed by Barbara Lewalski (1994), 'Milton and the Hartlib Circle: Educational Projects and Epic *Paideia*', in D.T. Benet and M. Lieb (eds), *Literary Milton: Text, Pretext, Context*, Pittsburgh, PA: Duquesne University Press, pp. 202–19.

2. Bacon, F. (1858), *Novum Organum Book II*, in Spedding, J. (ed.), *The Works of Francis Bacon*, ed. J. Spedding et al., London: Longman, vol. 4, p. 248.
3. Winstanley, G. (1649) *The True Levellers Standard Advanced*, in Sabine, G.H. (ed.) (1965), *The Works of Gerrard Winstanley*, New York: Russell & Russell, pp. 247–66.
4. Ibid., p. 253.
5. Ibid., p. 265.
6. More, T. ([1516] 1992) *Utopia*, London: Donald Campbell, pp. 23–4.
7. Thirsk, J. (1970), 'Seventeenth-Century Agriculture and Social Change', *The Agricultural History Review*, **18**, 148–77.
8. McRae, A. (1992), 'Husbandry Manuals and the Language of Agrarian Improvement', in M. Leslie and T. Raylor (eds), *Culture and Cultivation in Early Modern England: Writing and the Land*, London: Leicester University Press, pp. 35–62.
9. On Bacon's influence on agrarian and artisan literature of this period, see Haber, F.C. (1986), 'Time, Technology, Religion, and Productivity Values in Early Modern Europe', in F.C. Haber, *Time, Science, and Society in China and the West*, Amherst, MA: University of Massachusetts Press, pp. 79–92.
10. Webster, C. (ed.) (1974), *The Intellectual Revolution of the Seventeenth Century*, London: Routledge & Kegan Paul, p. 4.
11. Webster, C. (ed.) (1970), *Samuel Hartlib and the Advancement of Learning*, Cambridge: Cambridge University Press, pp. 6–8.
12. Milton, J. ([1644] 1953), 'Of Education. To Samuel Hartlib', in *Complete Prose Works*, 2 vols, New Haven and London: Yale University Press, vol. 2, p. 361. Despite Milton's evident awareness of Comenian philosophy, Barbara Lewalski (1994) assumes that he withheld active support for Hartlib's projects. For evidence of collaboration between Hartlib and Milton during the publication of the latter's *Of Education*, see Raylor, T. (1993), 'New Light on Milton and Hartlib', *Milton Quarterly*, **27** (1), 19–31.
13. Webster (1970), p. 2.
14. On the Hartlib Papers Project database at Sheffield University, see Greengrass, M. (1993), 'Interfacing Samuel Hartlib', *History Today*, **433**, 45–9; also the Introduction to M. Greengrass, M. Leslie and Raylor T. (eds) (1994), *Samuel Hartlib and the Universal Reformation*, Cambridge: University Press, pp. 1–25.
15. The significance of Bacon's thinking within the century's increasing division between mind and matter is discussed by Timothy Reiss in Reiss, T. (1982), *The Discourse of Modernism*, Ithaca: Cornell University Press, p. 95.
16. *Instauratio Magna*, possibly the most controversial and shadowy of Bacon's works, is outlined in Bush, D. (ed.) (1962), *English Literature in the Earlier Seventeenth Century 1600–1660*, Oxford: Clarendon Press, pp. 275–6. On the Christian eschatological underpinning of Bacon's philosophical programme, see Webster (1975), pp. 22–3.
17. Reiss (1982), pp. 181–4.
18. Clucas, S. (1994), 'In search of "The True Logick": Methodological Eclecticism

among the "Baconian reformers"', in M. Greengrass, M. Leslie and T. Raylor (eds), *Hartlib and Universal Reformation*, Cambridge: Cambridge University Press, pp. 51–74. On Comenius and the relation of *pansophia* to Baconian precepts, see Capkova, D. (1994), 'Comenius and his Ideals: Escape from the Labyrinth', in Greengrass et al. (1994), pp. 75–91.

19. [Plattes, Gabriel] *A Description of the Famous Kingdom of Macaria* (1641), repr. in Webster (1970), pp. 79–90.
20. Webster (1974), p. 377; and on Plattes, pp. 373–9.
21. *Macaria* (1641), in Webster (1970), p. 81.
22. Ibid., p. 82.
23. Ibid., p. 87.
24. Ibid., p. 89.
25. Ibid., p. 86.
26. Raylor, T. (1992), 'Samuel Hartlib and the Commonwealth of Bees', in M. Leslie and T. Raylor (eds.), *Culture and Cultivation in Early Modern England*, London: Leicester University Press, pp. 91–129.
27. Hartlib, S. (1655), *The Reformed Commonwealth of Bees. Presented in Several Letters and Observations ... With The Reformed Virginian Silkworm*, London: Giles Calvert.
28. Hartlib (1655), p. 41.
29. Ibid.
30. Ibid.
31. Ibid., pp. 9, 40, 49.
32. Foucault, M. (1970), *The Order of Things: An Archaeology of the Human Sciences*, London: Routledge, pp. 21–3, 51; Raylor (1992), p. 117.
33. In 1714 and the 1720s–30s, the lines were reprinted by Bernard Mandeville; see Mandeville, B. (1714: 1924), *The Fable of the Bees; or, Private Vices Publick Benefits*, ed. F.B. Kaye, **1**, Oxford: Clarendon Press, p. 24.
34. Raylor (1992), pp. 117–18. The very power and directness of Mandeville's language caused attacks from readers who misunderstood the subtlety of his dualistic reasoning: see Mandeville, ([1714] 1924), pp. xxxviii–lxxvi.
35. Thomson, J. ([1746] 1972), 'Autumn', in J. Thomson, *The Seasons*, ed. J. Sambrook, Oxford: Clarendon Press, pp. 120–21, ll. 1172–207.
36. Clucas (1994), p. 73.
37. Purver, M. (1967), *The Royal Society: Concept and Creation*, London: Routledge and Kegan Paul, esp. the 'Introduction' by H.R. Trevor-Roper (pp. xi–xvii); and pp. 207–34. See also Trevor-Roper, H. (1967), 'Three Foreigners: The Philosophy of the Puritan Revolution', in *Religion, Reformation and Social Change*, London: Macmillan, pp. 237–93.
38. Purver (1967), pp. 196–205. In rebuttal, see Webster (1975, p. 58) where the author reviews the evidence that 'during 1646, when the Invisible College was formed, Boyle was not acquainted with Hartlib'; the two men first met the following year.
39. Hall, A.R. (1963), *From Galileo to Newton*, London: Collins, pp. 141–2; *The Cambridge Economic History of Europe* (1967), London: Cambridge University Press, vol. 4, pp. 118–19.
40. Foucault (1970), p. 54.
41. A description of these interrelated systems can be found in Prochazka, M. (1995), 'Education and Exchange: Comenius and Samuel Hartlib', *Literaria Pragensia: Studies in Literature and Culture*, **5** (9), 32–7.

42. Ibid., p. 33, my italics.
43. Webster (1974), p. 47.
44. Hobbes, T. (1991), *Leviathan*, ed. R. Tuck, Cambridge: Cambridge University Press, pp. 287–96; Pocock, J.G.A. (1970), 'Time, History and Eschatology in the Thought of Thomas Hobbes', in J.H. Elliott and H.G. Koenigsberger (eds), *The Diversity of History: Essays in Honour of Sir Herbert Butterfield*, London: Routledge & Kegan Paul, pp. 149–98.
45. Hobbes (1991), pp. 286 and 287–96.
46. Pocock (1970), pp. 163–7.
47. Hobbes (1991), p. 310.
48. Pocock (1970), p. 175.
49. Prochazka (1995), pp. 34–5.

PART III
GENDER, SEXUALITY AND SCIENTIFIC THOUGHT

CHAPTER EIGHT

Journeys Beyond Frontiers: Knowledge, Subjectivity and Outer Space in Margaret Cavendish's *The Blazing World* (1666)

Bronwen Price

There is good reason to be interested in the work of Margaret Cavendish at the present moment, for her writing raises many of the questions which have concerned English Studies in recent years. In what ways does women's writing alter our view of the early-modern period? What kind of interventions with other writings of the period does it provide? For example, what sort of configurations of subjectivity and sexuality does Cavendish's writing offer?

These questions form the substance of this essay. Much of Cavendish's work is concerned with contemporary explorations into the new regions of science. However, although married to Sir William Cavendish, an important patron of natural philosophy with his own scientific circle, her access to such areas of study was limited, primarily because of her gender. She was not permitted to become a member of the Royal Society – the first formal academy of science, set up in 1660 – and was frequently derided, her interests being viewed as excessive. Her husband's attitude towards her philosophical endeavours seems to have been equivocal. Unusually for the period, he supported the publication of her work, yet appears to have distanced her from his manuscript coterie and, while Margaret Cavendish met Hobbes, Descartes and Gassendi, she remained only on the margins of Sir William's scientific circle. Her encounters with Descartes were indeed most unfruitful, for 'he did appear to me a man of the fewest words I ever heard'.[1]

Nevertheless, Cavendish wrote extensively about the very subjects from which she was barred. Nowhere is this more apparent than in her utopian fantasy *The Blazing World* (1666), a work which highlights the concept of shifting frontiers. *The Blazing World* does not address contemporary ideas simply in a straightforward way; it continually requires us to readjust our perspective, disconcertingly alternating between radical and more apparently conservative positions.[2] The first section of this essay concerns the narrative strategies *The Blazing World* employs to produce such an effect; the second explores its interventions in scientific debates. The relationship between the representation of politics, religion and gender is investigated in the third section, while the fourth examines various figurations of subjectivity. The final section considers the outer regions of the work, the prologue and epilogue, and their relation to the central narrative. Each of these areas is underscored by the idea of traversing boundaries, a motif to which *The Blazing World* continually returns.

The narrative

The significance of boundary-crossing is apparent in the narrative structure itself. Indeed, one of the most striking features of the narrative is the way it keeps transforming shape. Cavendish's preface 'To the Reader' itself notes this modulation between what is described as *The Blazing World*'s 'romancical', 'philosophical' and 'fantastical' parts. This is demonstrated at the beginning of the narrative, which bears the hallmarks of a romance. It opens with the abduction of an unnamed 'young lady' by a merchant who 'steals' her away by sea in order to obtain his 'desire'.[3] We might assume that this act of 'theft' will be followed by a rescue operation in which the lady will be repossessed and returned to her native country and class. In such a familiar scenario men are the agents of exchange and women commodities to be circulated in the interests of what Irigaray describes as '[t]he economy of desire – of exchange' which 'is man's business'.[4] However, in *The Blazing World* the abductor's boat is set off course by a storm and this marks a change of course in the narrative.

While the merchant seems initially to set the terms of the story, the tempest results in the male crew not knowing 'what to do or whither to steer their course'.[5] This signals the abductor's imminent loss of possession of the woman and control over her story, prefiguring the shipwreck at the North Pole and the death of all the men. What follows does not entail the lady's reappropriation within a 'legitimate' masculine economy of desire, but a relocation in the narrative's direction, which exceeds the boundaries of conventional romance. The boat is forced into another world, the Blazing World, which joins the lady's native world at its pole. On her entry into this new domain the lady is metamorphosed from a stolen object into an Empress, who presides over a male community, expresses desires of her own, finds pleasure in a platonic female relationship and eventually rescues her own world from war. She enters a space outside the frontiers of her native country, where the terms of that world and the narrative established within it are decentred, viewed from a world ex-centric to it. The voyage into the Blazing World enables us to see such conditions of sexual exchange differently, allowing for an alternative set of possibilities to those we might expect.

The opening of *The Blazing World* thus presents two types of narrative which change direction and alter shape through crossing boundaries with each other. As we travel from one type of narrative into another overlaying one, so our perspective is disoriented and repositioned. This 'many-in-one' feature is echoed throughout the text, resulting in a dialogue between one way of shaping understanding and another which serves to challenge emerging concepts of knowledge, subjectivity and sexuality.

With reference to this point it is worth noting the broader textual context of *The Blazing World*. The work was originally appended to Cavendish's *Observations Upon Experimental Philosophy*, which provides a critique of Robert Hooke's *Micrographia* (1665) and the empirical paradigm generally. Cavendish's preface to *The Blazing World* pinpoints the distinction between her two texts. The 'philosophical contemplations' of *Observations* are regarded as 'serious', being founded on 'reason' and enquiry after 'truth', while *The Blazing World* is regarded as 'a work of fancy', which creates 'of its own accord whatsoever it pleases and delights in its own work'. *Observations* is thus represented as the central text to which *The Blazing World* is the inauthentic adjunct, forming a diversion from 'more serious contemplations'. As woman is traditionally identified as the supplementary term to male primacy in cultural constructions of gender difference, so fiction is to natural philosophy here.[6]

However, though set in opposition, *Observations* and *The Blazing World* are joined together as 'two worlds at the ends of their poles'. Thus connected, the works are presented as comprising neither one nor two texts. They are differentiated in terms of their modes of representation, but intersect at their axes; together, the one casts a new light on the other to produce a conversation between them. Indeed, philosophical science is regarded as sometimes requiring 'the help of fancy to recreate the mind'.[7] In her illuminating discussion about *The Blazing World*'s intellectual contexts, Rosemary Kegl takes this idea further. Arguing that 'the joining of "two Worlds at the ends of their Poles"' is Cavendish's 'governing metaphor' for her narrative of intellectual process, Kegl proposes that Cavendish thus suggests a means by which fancy is one of science's 'essential components'. In turn, *The Blazing World* demonstrates 'that women might acquire both the *concepts* and the contemplative *habits* that are central to intellectual life', most notably those of science.[8]

Moreover, in placing 'fancy' next to 'serious philosophical contemplations', Cavendish brings to bear on science those imaginings which from the seventeenth century onwards scientific discourse increasingly attempted to exclude: what was regarded as inauthentic, delusion, fantasy. That discourse is best exemplified by Francis Bacon (1561–1626), whose empirical, experimental method, taken up by figures such as Hooke, formed the basis of the Royal Society's scientific practice. Echoing Bacon's concern to establish what 'is sound and without mixture of fables',[9] Hooke argues that 'the Science of Nature has been already too long made only a work of the Brain and the Fancy: It is now high time that it should return to the plainness and soundness of Observations on material and obvious things.'[10]

The Blazing World, however, highlights the very areas which are repressed by these incipient divisions of scientific discourse. It is a fantasy, moreover, which brings to its centre a woman presiding over the range of epistemological fields. In bringing together 'science' and fiction, this apparently supplementary 'piece of fancy' problematises the boundaries within which concepts of gender and genre are formulated, for *The Blazing World* crosses paths and interacts with *Observations*, creating a world which unsettles the social circumscriptions identified in the latter work. Indeed, the prefaces to *Observations* both draw attention to women's exclusion from learning, 'as being not suffer'd to be instructed in Schools and Universities', and project that the speaker's own entry into the arena of natural philosophy will be viewed as 'an extravagant, or at least a Fantastical disease'.[11] *The Blazing World*, however, re-examines *Observations*' attacks on the new experimental, empirical science, locating them in a new context of different possibilities.

Science

Shortly after the lady is made Empress of the Blazing World, she founds specialist schools in arts and sciences, of which all the inhabitants become members according to their particular expertise. She begins her investigations by summoning the birdmen and commanding 'them to give her a true relation of … the sun and moon'. In broad terms their findings resemble the recently established Copernican view of the universe, the sun being regarded as 'fixed and firm like a centre' around which the other planets revolve. Further, the 'two celestial bodies' are represented in traditional gendered terms, where the sun is male and the moon female, identified by 'her' difference and supplementariness to the primary 'masculine' term.[12] In Copernicus' writing the synonymity between the sun's centrality and male authority is articulated in an elaborate image of imperial power: 'In the middle of all sits the Sun enthroned … He is rightly called the Lamp, the Mind, the Ruler of the Universe … So the Sun sits as upon a royal throne ruling his children the planets which circle round him'.[13] Here the sun's representation as centre of the universe and source of light is employed as a means of naturalising patriarchal power. The same equation may be inferred between masculine authority and the 'extraordinary splendour' of the sun, which is 'like a centre', in the birdmen's observations.[14]

However, the portrayal of the 'female' moon disrupts such notions, for in Cavendish's text the moon does not circle round the sun. It has its own centre, but is at the same time 'excentrical' to the solar system, thus problematising what a centre might be. While the moon inhabits the same sphere as the sun, it is not easily contained within it, possessing its own source of light, and so shedding another light, another perspective, on the solar system.[15] The birdmen's description of the sun as being 'like a centre' seems to privilege one set of possibilities while repressing others, something which is stressed by the narrator's interventions. For her the sun is not 'fixed and firm like a centre', but rather 'a globous fluid body' with 'a swift circular motion'.[16] She thereby shifts our point of perspective, asking us to reconsider how it is interpreted.

The question of how things are given shape and its relation to the formulation of gender, power and knowledge are, as Evelyn Fox Keller observes, especially significant to the discourses of early-modern science.[17] Both Keller and Merchant suggest that Bacon's writing represents a key instance of a rhetoric in which nature is sited as a feminised object to be discovered, known and mastered by a rigorous masculine mind.[18] It is thus particularly noteworthy that *The Blazing World* contains significant parallels to Bacon's fragmentary fictional work, the *New Atlantis* (1627). The *New Atlantis* is similarly appended to a volume of natural philosophy, *Sylva Sylvarum*, and the story begins in an analogous way with a sea voyage of a male crew who are set off course by a storm that carries them north. While expecting to perish, the crew, like the Empress, are rescued by divine intervention. However, while the Empress arrives at a new world where she sets up her own scientific community, Bacon's protagonists land at the island of Bensalem which, though unknown, is contained within the frame of the Earth. There they are presented with Bensalem's already established scientific centre, Salomon's House, by one of its governors. This is a foundation 'dedicated to the study of the Works and Creatures of God',[19] divided into specialist areas of scientific study and, like the Empress's scientific body, designed to 'be beneficial to the public'.[20]

The ideas informing each community, however, differ in important ways. In many respects Salomon's House exemplifies a set of scientific practices proposed by Bacon in his earlier works, such as *The Advancement of Learning* (1605), *The Great Instauration* and *Novum organum* (both 1620). While conceding that 'nature to be commanded must be obeyed', these works are underscored by a sexual political rhetoric in which 'the secrets of nature' are to be disclosed by 'penetrating' 'her' 'inner chambers' so as to 'extend the power and dominion of the human race itself over the universe'. Nature is thus required to 'give herself into' the 'hands' of men.[21] In the *New Atlantis* these images of conquest are located in the context of a fully operative 'prototype … research institute'[22] that dedicates itself to 'penetrating into nature' so as to establish 'the knowledge of Causes, and secret motions of things; and the enlarging of the bounds of Human Empire, to the effecting of all things possible'.[23]

The Blazing World, however, troubles the grounds of such practices. By their very constitution the Empress's virtuosi destabilise Bacon's differentiation between knower and known, combining both man and beast. They are identified as being integral to their very areas of enquiry – the birdmen explore the atmosphere; the fish and mermen, marine life; the worm-men resemble the maggots they study. The object of knowledge is not something which is extractable from those who examine it, but merges with them. The hybridisation of the scientists themselves unsettles the processes of identification and classification with which they seem to be engaged and worries Bacon's concept of putting 'nature under constraint', and having 'her' 'squeezed and moulded'.[24]

Moreover, while the assembly of the virtuosi appears to offer an extensive compilation of scientific data, this section of *The Blazing World* also places in continual question what constitutes knowledge and the methods by which it is obtained. In responding to the Empress's request to know 'what kind of substance or creature the air was', the birdmen highlight the limits of empirical enquiry: (for nature is so full of variety that our weak senses cannot perceive all the various sorts of her creatures, neither is there any one object perceptible by all our senses, no more than several objects are by one sense).[25]

Like the Empress's other 'new-found societies', the birdmen's observations offer a particularised area of knowledge which resembles the specialised fields of expertise represented in Bacon's *New Atlantis*. However, while Salomon's House represents the source of secret knowledge exclusive to its members, signalling their access to 'the secrets of nature', the birdmen regard the natural world as unfathomable. Their enquiries question what kind of things can be known and the premises upon which identities are shaped. One way of forming understanding is made contingent through its contiguity with another. Nature is thus transformed from a domain which we think we know – experienced through our deficient senses; as, for example, air and wind – into a world beyond the frontiers of our comprehension.[26]

The question of empirical investigation comes to the foreground again when the Empress commands the bearmen, her experimental philosophers, to observe celestial bodies 'through such instruments as are called telescopes'.[27] In Bacon's experimental method, the use of optical aids is recommended as a means of extending and refining the visual faculty, offering 'substitutes to supply its failures'.[28] In *New Atlantis*, for example, the members of Salomon's House have 'procure[d] means of seeing objects afar off; as in the heaven and remote places'.[29] Similarly, Hooke's *Micrographia* refers to 'the invention of Optical Glasses' as 'supplying' the 'infirmities' of the senses.[30] In *The Blazing World*, however, the telescope is presented as producing multiple points of perspective, various mappings of the planets, which simultaneously cancel each other out. The telescope's standard, collective eye paradoxically has the effect of fragmenting and decentring the clear line of vision sought by Bacon and Hooke; for outer space emerges as that which cannot be sighted/sited in a distinct shape or form. Instead, the telescopes are 'false informers' which produce 'artificial delusions',[31] precisely echoing the terms that Bacon tries to 'exorcise' from his scientific method.[32]

The limitations of such 'exquisite instruments'[33] are interrogated once more when the Empress examines various bodies under 'several other artificial optic glasses', including microscopes, 'by the means of which they could enlarge the shapes of little bodies, and make a louse appear as big as an elephant'.[34] In *Micrographia* Hooke suggests that 'by the help of Microscopes, there is nothing so small, as to escape our inquiry; hence there is a new visible World discovered to the understanding'.[35] But while the bearmen claim that the microscopes 'rectify and inform their senses', the narrator notes: 'Only this was very remarkable … that notwithstanding their great skill, industry and ingenuity in experimental philosophy, they could yet by no means contrive such glasses by the help of which they could spy out a vacuum, with all its dimensions.'[36] While Hooke envisages the microscope as the means through which 'we may perhaps be inabled to discern all the secret workings of Nature',[37] the narrator reflects on the vacant spaces of scientific investigation, areas of intelligence that such 'optical glasses' cannot 'spy out'. She contemplates phenomena that lie in the outer recesses of existence; unidentifiable fluid objects – 'immaterial substances, non-beings, and mixed-beings, or such as are between something and nothing' – which escape incorporation into processes of classification and identification. Clear demarcations between presence and absence, being and non-being, certainty and doubt are thereby destabilised.[38]

The Empress's relentless questions thus result not so much in an accumulation of knowledge, but rather slide towards an arena of other possibilities, which refuse to 'overcome' 'the difficulties and obscurities of nature'.[39] Indeed, her satisfaction with the replies she receives derives from a sense of what is beyond comprehension. The most assured conclusion that is reached about nature is its impenetrability: it has no point of origin or closure, but comprises many-in-one, existing in a constant state of flux, forever reforming and transforming itself.[40]

Absolutism and gender

If this begins to sound as if nature has ousted God as the source of an unfathomable, mysterious knowledge, our perspective is readjusted when, at the scientific assembly's close, the Empress rejects the logicians' arguments by alluding to a transcendent, divine order:

> I do believe that it is with natural philosophy as it is with all other effects of nature, for no particular knowledge can be perfect … nature herself cannot boast of any perfection but God himself, because there are so many irregular motions in nature, and 'tis folly to think that art should be able to regulate them, since art itself is for the most part irregular. But as for improbable truth, I know not what your meaning is … there is so much difference between truth and improbability that I cannot conceive it possible how they can be joined together.[41]

While previously regarded as 'eternal and infinite', nature is now underwritten by the Word, contained within the law of God the Father. The apparently self-determining, 'self-moving' principles of nature are held in check, it seems, by the prime mover, in reference to whose perfection nature's 'perpetual changes and transmutations' are defective. John Rogers notes how Cavendish's view 'that nature is not only autonomous but fully conscious' presented in her natural philosophical works provoked a vehement response through its 'threat to the belief in the omnipotence of God'.[42] Here, however, it seems that ultimately God's mysteries, rather than nature's, deny access to 'the knowledge of nature', for, as the spirits later state, 'not any creature but God himself can have an absolute and perfect knowledge of all things'.[43] Knowledge is thus conterminous with secrets, secrets which maintain a 'privileged and protected status' through their fundamental link with God.[44] Having ventured to the frontiers of secularism, the text seems unable to go over the edge into a different world of understanding, but instead retreats back into a familiar sphere of pre-modern thought in which God is the ultimate guardian and principle of knowledge.

The logicians' assertions about improbable truths thus cause the Empress considerable anxiety, for, if truth may be improbable, God's place as the source of absolute truth is also rendered uncertain. Further, if God's position is troubled, so, too, are the social and political fabric over which she governs with 'an absolute power', conferred on her by the Emperor's divine right.[45] The effect of the logicians' syllogisms puts her 'brain on the rack', the image of internal disquiet bringing to mind anxieties about disturbances in the body politic over which she presides as symbolic head. She thus decides to 'confine' the logicians' 'disputations to your schools … lest they disturb also divinity and policy, religion and laws'.[46]

There is, indeed, throughout the text an underlying worry about civil strife, together with a concern to promote the ideology of absolutist monarchy. When the Empress arrives at the imperial city of the Blazing World she learns 'there was but one language in all that world, nor no more but one Emperor', 'for as there is but one God, whom we all unanimously adore with one faith, so we are resolved to have but one Emperor, to whom we all submit with one obedience'.[47] Monarchy is perceived in terms of mid-seventeenth-century royalist iconography, the emblematic figure of the sovereign being presented as the pivotal sign which binds all its members together in an inclusive, organic, unitary frame.[48] The King is represented as head of the body of state of which he is ruler and father, and this parallels his role as agent of God, the ultimate *paterfamilias*. Harmony within the monarchic anatomy reproduces those features elsewhere: the subjects' obedience to the King corresponds to, is reinforced by and reinforces their submission to God.

However, as already noted, *The Blazing World* has a habit of appearing to be one thing while simultaneously turning into something else. Ironically, the Empress's entry into the imperial city both produces and discloses plurality, breaking open the enclosed, unitary frame that absolutist monarchy attempts to contain. Once she is granted 'an absolute power to rule and govern all that world as she pleased',[49] the plenary figure of the Emperor seems to be displaced and transformed. He disappears from the text until the second part of *The Blazing World*, while the Empress opens up a range of other possibilities, which she herself acknowledges may produce dissension.[50]

Rogers importantly notes that, although Cavendish never questions 'the ideology of aristocratic privilege', by refusing to admit 'any single regulatory power or absolute centre of command' and by asserting a 'disseminated sovereignty' in her natural philosophical works, she invokes a Nonconformist rhetoric. This rhetoric is employed, he argues, implicitly to oppose not the tyranny of religious or political institutions, but that of sexual subjection. Such an anti-authoritarian perspective is apparent in the decentred government of *The Blazing World* over which the Empress now presides.[51]

The Empress illuminates the repressed areas of the Blazing World's absolutism, exposing the system of differences upon which it rests. On enquiring about its religious practices, she learns that its 'men' have 'all but one opinion concerning the worship and adoration of God', but notices that women are barred from religious assemblies, discovering that they 'say their prayers by themselves in their closets'. This is justified by the priests because '"It is not fit," said they, "that men and women should be promiscuously together in time of religious worship, for their company ... makes many, instead of praying to God, direct their devotion to their mistresses"'.[52] This Paradise (the imperial city's name) is thus understood through 'knowledge' of the Fall. Women are already viewed as temptress figures, defined by their sexuality and its potential to cause disruption. Yet women are also grouped with children as being mischievous but implicitly not responsible for their actions. They thus possess a different kind of knowing from that of men, one that is intrinsically tied to their sexuality but also bound to 'ignorance'.

As such, they are not properly subjects, having 'no employment in church or state', merely residing at the edges of this absolutist patriarchal system.[53] However, Catherine Gallagher reveals that the women's exclusion from all public offices except for that of the monarch itself indicates that they are thereby not 'circumscribed always by the terms of political subjection'.[54] Both included and excluded, they represent the residual term upon which male centrality is premised, but by which it is also troubled. While their physical confinement is a means of controlling their sexuality, women's privacy also provides scope for more insidious forms of subversion; as the priests explain, 'many times … they cause as much, nay more mischief secretly, than if they had the management of public affairs'.[55] Women thus point to the latent anxieties of division and transgression which are initially hidden by the seemingly unitary bonds of absolutism, but which become increasingly apparent through the Empress's questions and decisions.

Finding the manner of the Blazing World's religion 'very defective' and its people's knowledge of 'the divine truth' deficient, the Empress converts them to her own religion and appoints herself head of the church which is to include 'a congregation of women'. This apparently reformist measure transforms what 'woman' signifies, for no longer are they constructed in reference to their sexuality, but through their intellectual agency, being found to possess 'quick wits, subtle conceptions, clear understandings and solid judgements'. From being isolated, feared and excluded, they are remodelled into a community of obedient subjects who 'became in a short time very devout and zealous sisters'.[56]

Ironically, then, the Empress's liberating act is encoded within what are themselves authoritarian confines which demand devotion to her, and to God through her. Having ruptured the boundaries of absolutism and released difference, she herself establishes frontiers which embody absolutism. She is now the pivot through whom 'the divine truth' may be realised.

Or is she? No sooner has the Empress apparently located a stable centre of meaning as 'the divine truth' to which she provides admission, than she summons the 'immaterial spirits' of the Blazing World to question them about theological matters and the Cabala's exegesis. What is most striking about this episode are the kinds of questions that the Empress asks and the furious pace at which they are delivered. Where is 'the Paradise'? she asks. Is it 'in the midst of the world as a centre of pleasure, or … the whole world or a peculiar world by itself'?[57] But having proposed these various possibilities she discovers from the spirits that she herself is in Paradise, which we already know is the imperial city's name. Identifying and locating paradise is therefore imbued with a sense of disorientation, as the Empress has not until now comprehended the nature of the place at which she has arrived.

Moreover, the Empress's questions about paradise highlight the narrative pattern as a whole, which both revisits and revises the original Biblical story. The year before Milton attempted to 'justify the ways of God to men' in *Paradise Lost*,[58] *The Blazing World* provides, it seems, an ironic reformulation of the paradise myth, presenting itself through the perspective not of God, nor of Adam, but of the Eve figure of the Empress. In this paradise regained it is a woman who is forced to leave her native country because of a man's misdemeanour. He, along with his collaborators, is punished for this act, while she is saved and rewarded, being sent to paradise and given free rei[g]n over it. She is thus liberated from being an object of male desire, releasing in the Empress an insatiable appetite for knowledge. She appears as the 'Bold … adventurous Eve',[59] venturing into terrains of knowledge from which in her own world she would be forbidden access. Without trepidation she explores the unquestionable, enquiring, amongst other things, whether the universe was made in six days, whether there are any atheists in the world, and where heaven and hell are. While claiming to be 'satisfied' with the spirits' reply that such matters are 'beyond your knowledge and understanding', the Empress nonetheless responds by asking more questions. In the second part of *The Blazing World*, this audacious Eve figure returns to the fallen world to which Adam fled and rescues it in apocalyptic form by a 'miraculous delivery'. While mortal, she appears in this later section as 'some celestial creature' who has the power 'to walk upon the waters and to destroy whatever she pleases'.[60] The Empress is metamorphosed into a Christ-figure or Second Eve.

And yet the Empress's story is still haunted by the original myth, as she repeatedly asks where the fault for the fall from paradise lies. These restive phantoms of the original story suggest the sense of denial out of which the Blazing World's paradise is formed, but signal the retelling that is taking place. In the Blazing World's newly-discovered paradise the Empress reshapes what she finds afresh, opening it up to question through her relentless inquisitiveness. Paradise is represented as a contestable space without finite boundaries, rewritten and reread for different possibilities.

If this is so with Paradise, so it is with 'the divine truth' within which paradise operates. Indeed, the Empress' sudden suppression of her questions for fear of wrongdoing at once instigates in her 'a great desire' to 'make' her own Cabala,[61] a desire which echoes Eve's appetite for acquiring hidden, forbidden knowledge. The Empress's divine authority thus slides into transgression, the Absolute into possibility and improbability. No sooner have her questions been curbed than she produces further textual excess, opening up God's Word to negotiation.

Subjectivity

The Empress's desire to create a Cabala produces, however, yet another transformation in the text. In order to fulfil her enterprise she requires the aid of a spiritual scribe. Having discounted 'Galileo, Gassendus, Desartes, Helmont, Hobbes, H. More, etc.' on the spirits' advice that they were 'so self-conceited that they would scorn to be scribes to a woman', the Empress requests the help of the Duchess of Newcastle who, though 'not one of the most learned, eloquent, witty and ingenious', is recommended as 'a plain and rational writer' whose works possess 'sense and reason'.[62] The Duchess of Newcastle is thus inscribed into her own narrative as a soul which travels beyond the realms of Britain to the Empress in the Blazing World.

Ironically, it is the Empress's summoning of the Duchess as her scribe which revises her plans, for the Duchess dissuades her from writing a Cabala. Rather, the fulfilment of the Empress's 'great desire' is transposed by the platonic relationship opened up by the Duchess's entry into her soul. Unlike the conventional platonic female lover, however, the Duchess is not the mere bodily instrument which provides her suitor access to an ideal divine truth; instead, she generates an active extra-terrestrial female exchange which eludes such sitings/sightings of woman. The union between the Duchess and the Empress takes place inside the latter's body: it cannot be seen, and yet involves 'no concealment' between them; it is indivisible, but embraces autonomy, as each 'lover' is able to journey within and out of the other's soul across the frontiers of body and space.[63]

Even before the arrival of the Duchess, we learn from the spirits that platonic lovers are 'not only very intimate and close, but subtle and insinuating'. In this sense, the platonic relationship between the women indicates a melting of boundaries between selves and a subsequent reshaping of identity. In relation to the Duchess, the Empress is neither clearly sovereign nor subject, one thing nor another, the Empress and Duchess 'being like several parts of one united body' – many and one.[64] While it appears that she commands the Duchess's actions in that she 'sent for the Duchess' who was 'ready to wait on' her, the Empress is nonetheless subject to the Duchess's representations. In particular, the Empress's 'great desire' to write a Cabala is displaced by the Duchess's own 'ambitious desire' to be like her, an Empress of a world – a desire which knows no bounds. It is the pursuit of this desire which crosses bounds with, and decentres, the narrative of the Empress, who now becomes facilitator of the Duchess's 'extreme ambition'.[65]

The Empress consults her immaterial spirits to advise the Duchess on how best to accomplish her wish. They, however, question the Duchess's aspiration 'to be Empress of a terrestrial world whenas you can create yourself a celestial world … within the compass of the head or skull'. Her desire is subsequently converted from the ambition to be sovereign of a world without to being that of one within, and is to be fulfilled, not by altering an external terrain through conquest and appropriation, but through the metamorphosis of the interior realm of the brain. The journey into outer space leads the Duchess across new frontiers within an inner place, inaugurating the possibility of creating 'a world of my own'.[66]

In certain ways the Duchess's newly-found identity as sovereign of her own inner world contains resonances of the emergent modern notion of an autonomous sense of self, to which the philosophy of René Descartes (1596–1650) is central. In contrast with Bacon's inductive method, Descartes privileges a deductive one based on a theory of rationalism, dualism and interiority, in which 'we perceive bodies only by the understanding which is in us ... we do not perceive them through seeing them or touching them, but only because we conceive them in thought'. Descartes thus asserts, 'I know clearly that there is nothing more easy for me to know than my own mind',[67] so that the mind is radically divided from the body. It is this inner domain of the thinking mind which defines the Cartesian subject and which, like the enclosed interior world of the Duchess's skull, is identified as the centre of knowledge and meaning.

However, while echoing Cartesian notions of subjectivity, the figure of the Duchess also dismantles them. Indeed, as with her questioning of Baconian epistemology, Cavendish's troubling of Cartesian rationalism is prevalent in her work generally.[68] Cartesian identity is affirmed through the assertion and defence of its boundaries, 'contain[ing] itself within the precise limits of truth' and excluding all that is confused.[69] However, when the Duchess tests out a world specifically based on Cartesian principles in the formation of her interior sovereign state, the activity of the Cartesian 'ethereal globules' in her brain directly results in irrationality and instability. Her mind becomes 'dizzy' and slips into an unconscious state, when she is almost put 'into a swoon, for her thoughts, by their constant tottering, did so stagger as if they had all been drunk'.[70]

Having experimented with a variety of models from 'the ancient philosophers' and 'the moderns' and finding that they procure only 'an horrible pain in her head', the Duchess resolves to 'make a world of her own invention'. This self-created world within both recalls and makes strange the components of Cartesian identity, rearranging them into a different constellation and so bringing to them a difference of view. Like Descartes' interior, autonomous, thinking mind, the Duchess's inner world is composed 'only of the rational' and is 'well ordered and wisely governed', but for her the rational slides into fantasy, order into licence, government into enfranchisement, so as to produce 'an imaginary world' which may embrace limitless possibilities.[71] The Duchess's mind is not in any sense 'one single and complete thing' which may be easily known,[72] but an inclusive, elusive sphere, which is 'so curious and full of variety ... that it cannot possibly be expressed by words'.[73] Its 'framing' does not involve containment within a thinking, doubting, essentialised 'I', but rather signals excess. Her interior realm is protean and in process, for the spirits advise the Duchess that she may alter such a world as often as she chooses.[74]

Furthermore, the Duchess's internal identity does not possess a single or finite point of origin, for her soul is still incorporated within the Empress, who 'In the meantime ... was also making and dissolving several worlds in her own mind'.[75] The interior sovereign states of both Duchess and Empress are thus separate and yet contain each other, so that neither is clearly at the centre, each being reoriented by its relation to the other. Singularity is represented as embracing plurality; identity, difference; centre, margins; separation, fluidity.

Outer space

These questions of identity are made more complex by the Duchess's double inscription, for the Duchess within the narrative is joined at the opposite poles of the story to the Duchess without in the prologue and epilogue. This figure of the outer regions of the text crosses boundaries with that of the inner story, but is not identical with her, treating 'the figure of honest Margaret Newcastle' as an object of her narrative.[76] The Duchess within is written into the utopian 'no place' of the Blazing World where, though an alien, she is enabled to fulfil her 'ambitious desire'. The Duchess without, however, is historically situated and must negotiate the boundaries of a culturally defined gender identity in order to reconstruct herself into something different. She reminds us of the frontiers beyond which the fantasy of *The Blazing World* takes us and so sheds a different light on the central text.

In the context of her cultural and historical positioning, this figure of the Duchess is indeed a displaced person, an alien, who must go beyond authorised bounds in order to fulfil her ambition 'not only to be Empress but authoress of a whole world'.[77] Such a task involves a continual shifting between different subject positions. At one level she asserts her self-determination in the attainment of her place as Margaret the First, claiming boldly to go where no one, woman or man, has ventured before by creating a 'new world' in her head and presenting it to public view. Yet, in displaying this outer space within, she becomes subject to the reader's judgements: 'nobody, I hope, will blame me', she adds after her initial self-affirmation. If so, 'I must be content to live a melancholy life in my own world'.[78] Moreover, while she invites the readers to become subjects of her textual domain, she is also troubled with the anxiety of usurpation and the remapping of her newly authored territory.[79]

In this sense, the outer regions of the text highlight the politics of speaking, writing and knowing, presenting Cavendish's world within as a place in which a struggle for meanings and positions is played out. Such a struggle both calls up and contrasts with the military enterprises of the masculine world, whose assertion is represented in the fiction as involving the acquisition and annihilation of lands, property and people. Her own imperial project, however, requires no 'such disturbances'; she hopes to share her created wealth with her 'noble female friends' and presents her empire as a non-terrestrial space of textual fruition, which is open to continual reproduction.[80] The prologue and epilogue thus provide an interrogative border between fiction and non-fiction, place and no place, materiality and possibility, inviting the reader to look beyond the frontiers of the text and examine the world afresh from a different polar region.

Notes

1. Cavendish, M. (1655), 'An Epilogue to my Physical Opinions', in *The Philosophical and Physical Opinions*, London: F. Martin and F. Allestrye (unpaginated).
2. For a recent illuminating account of *The Blazing World*'s incorporation of radical and conservative stances, see Kegl, R. (1996), '"The world I have made": Margaret Cavendish, feminism and the *Blazing World*', in V. Traub, M.L. Kaplan and D. Callaghan (eds), *Feminist Readings of Early Modern Culture*, Cambridge: Cambridge University Press, pp. 119–41.
3. Cavendish, M. ([1666] 1991), *The Blazing World*, in P. Salzman (ed.), *An Anthology of Seventeenth-Century Fiction*, Oxford and New York: Oxford University Press, pp. 252–3.
4. Irigaray, L. (1985), *This Sex Which Is Not One*, trans. C. Porter, Ithaca: Cornell University Press, pp. 187–8.
5. Cavendish ([1666] 1991), p. 253.
6. Ibid., pp. 251–2. Cavendish's preface to natural philosophers in *Poems and Fancies* (1653) makes a similar distinction between poetry and science. See Price, B. (1996), 'Feminine Modes of Knowing and Scientific Enquiry: Margaret Cavendish's Poetry as Case Study', in H. Wilcox (ed.), *Women and Literature in Britain 1500–1700*, Cambridge: Cambridge University Press, p. 126.
7. Cavendish ([1666] 1991), p. 252.
8. Kegl (1996), pp. 126–7. However, Kegl provides a more complicated account of this metaphor through which she argues Cavendish figures women's intellectual equality with men by exploring its use in the representation of merchant's capital, female desire and racial difference in *The Blazing World* (pp. 119–41).
9. Bacon, F. (1857–74), *The Works*, ed. J. Spedding, R.L. Ellis and D.D. Heath, 14 vols, London: Longman, vol. 4, p. 30.
10. Hooke, R. ([1665] 1961), *Micrographia*, preface R.T. Gunther, New York: Dover; Hooke's Preface does not contain page numbers. Hooke was an advocate of Baconian science and a key member of the Royal Society, being appointed Curator of Experiments in 1662 and becoming its secretary in 1677.

11. Cavendish, M. (1666) 'To the Most Famous University of Cambridge' and 'The Preface to the Ensuing Treatise', in *Observations Upon Experimental Philosophy*, London: F. Martin and F. Alleystre, no page numbers.
12. Cavendish ([1966] 1991), pp. 263–4.
13. Copernicus, N. (1543), *On the Revolution of the Celestial Spheres*; cited in T.S. Kuhn (1957), *The Copernican Revolution*, Cambridge, MA, and London: Harvard University Press, pp. 179–80.
14. Cavendish ([1666] 1991), pp. 263–64.
15. Idem.
16. Ibid., p. 264.
17. Keller, E.F. (1992), *Secrets of Life, Secrets of Death: Essays on Language, Gender and Science*, New York and London: Routledge, pp. 56–72.
18. See Merchant, C. (1980), *The Death of Nature*, San Francisco and London: Harper and Row, pp. 164–90. For a more sophisticated account of this gendered rhetoric, see Keller (1992), p. 57, and Keller, E.F. (1985), *Reflections on Gender and Science*, New Haven and London: Yale University Press, pp. 33–42. I discuss Baconian discourse further in my essay on Cavendish's poetry (Price, 1996, pp. 120 and 129).
19. Bacon (1857–74), vol. 3, p. 145.
20. Cavendish ([1666] 1991), p. 281.
21. Bacon (1857–74), vol. 4, pp. 47, 28–29. Hooke, ([1665] 1961), in The Preface takes up Bacon's metaphors of sexual mastery, representing nature as an unruly woman, who, in order to be properly known, requires the most vigilant form of surveillance.
22. Merchant (1980), p. 182.
23. Bacon (1857–74), vol. 3, pp. 165 and 156. For an intriguing alternative reading of gendered discourse in *New Atlantis*, see Aughterson, K. (forthcoming, 2002), '"Strange things so probably told": gender, sexual difference and knowledge in *New Atlantis*', in B. Price (ed.), *Francis Bacon's The New Atlantis*, Manchester: Manchester University Press.
24. Bacon (1857–74), vol. 4, p. 29.
25. Cavendish ([1666] 1991), pp. 265–6.
26. Cavendish's troubling of empiricism in general is well documented. See, for example, Bowerbank, S. (1984), 'The spider's delight: Margaret Cavendish and the "female" imagination', *English Literary Renaissance*, **14**, 392–408, and Sarasohn, L.T. (1984), 'A science turned upside down: feminism and the natural philosophy of Margaret Cavendish', *Huntingdon Library Quarterly*, **47**, 289–307. Compare with Cavendish's *Grounds of Natural Philosophy* (1668), London: A. Maxwell, p. 163, in which she argues that 'there are different Knowledges, in different Creatures; yet, none can be said to be least knowing, or most knowing: for, there is (in my opinion) no such thing as least and most, in Nature.' She also makes similar points in her poetry, though from a less formulated perspective (see Price, 1996, pp. 128–30). Cavendish is, of course, not alone in raising questions about the value of empiricism during this period. Both Descartes and Hobbes favour a deductive method over Bacon's inductive one and, overall, Hobbes opposed the methods of the Royal Society (see Martin, J. (1991), 'Natural philosophy and its public concerns', in S. Pumfrey, P.L. Rossi and M. Slawinski (eds), *Science, Culture and Popular Belief in Renaissance Europe*, Manchester: Manchester University Press, p. 115). In general, the Empress's interest in the underlying causes that produce particular effects, such as the wind, parallels

Hobbes's method in *De Corpore* (1655). For a detailed account of the links between Hobbes and Cavendish, see Hutton, S. (1997), 'In Dialogue with Thomas Hobbes: Margaret Cavendish's natural philosophy', Women's Writing, **4** (3), 421–32. The crucial departures Cavendish makes from Hobbes's mechanistic, materialist philosophy in her later vitalistic work are, however, pointed out in Rogers, J. (1996), *The Matter of Revolution: Science, Poetry and Politics in the Age of Milton*, Ithaca: Cornell University Press, pp. 177–211.

27. Cavendish ([1666] 1991), pp. 266–7.
28. Bacon (1857–74) vol. 4, p. 26.
29. Ibid., p. 162.
30. Hooke ([1665] 1961), The Preface (no paginated numbers).
31. Cavendish ([1666] 1991), pp. 268–9.
32. Bacon (1857–74), vol. 4, p. 30.
33. Ibid., p. 26.
34. Cavendish ([1666] 1991), p. 269.
35. Hooke ([1665] 1961), The Preface.
36. Cavendish ([1666] 1991), p. 272.
37. Hooke ([1665] 1961), The Preface.
38. Cavendish ([1666] 1991), p. 272. In *Observations* (1666), however, Cavendish argues that the existence of vacuums is 'absurd' and 'impossible' (pp. 141–2), just as Bacon does in *Novum organum* (1857–74, vol. 4, p. 126). Cavendish's argument, though, rests on the idea that 'Nature is a body of continued infiniteness, without any holes or vacuities' (1666, p. 137), whereas Bacon takes a materialist line, wishing to 'be led only to real particles, such as really exist' (1857–74, vol. 4, p. 126). In *Grounds* (1668) Cavendish also considers the existence of inaccessible life forms, but 'cannot imagine ... there is any thing between Something and Nothing' (pp. 234 and 237). See also my discussion about Cavendish's speculation on the invisible in her verse (Price, 1996, pp. 127–9) and see pp. 38–9 above.
39. Bacon (1857–74), vol. 4, p. 314. The significance of ignorance to knowledge, which is such a central concept in *The Blazing World*, is apparent elsewhere in Cavendish's work, as I show in reference to her verse (Price, 1996, pp. 117–39). See also Cavendish (1655), pp. 2, 52 and 97–8, and (1668), p. 217.
40. This assertion about nature is reiterated throughout the scientific debate (see pp. 275 and 283 in *The Blazing World*), and is pre-empted in *Observations* (1666) which argues that 'Nature is one infinite body ... producing infinite effects' (pp. 47–8). In his fascinating account of Cavendish's shift towards vitalistic theories of nature in her later works, John Rogers (1996) compellingly argues that, like other writers of the period, Cavendish 'marshal[s] images from natural philosophy as an organisational foundation for her beliefs about human society and, more specifically, about the interaction of men and women' (p. 185). Rogers suggests that Cavendish's view of the self-determining movement and agency of nature provides an alternative to, and critique of, the 'masculinist ethics' of Hobbes's 'authoritarian mechanism' and 'rule of force'. Moreover, because she diminishes 'the empirical faith in the primacy of external force', Cavendish's vitalistic theory focuses 'instead on the much subtler interaction of the "parts" of matter that dwell *within* each material body' (pp. 185–90). Not only is this concentration on internal interaction apparent in *The Blazing World*'s representation of nature, but also in its social dynamic, as we shall see.
41. Cavendish ([1666] 1991), p. 288.

42. Rogers (1996), p. 194. Rogers particularly notes Ralph Cudworth's reactions.
43. Cavendish ([1666] 1991), pp. 281 and 295.
44. See Keller (1992), who argues that there was a 'rhetorical shift in the locus of essential secrets from God to Nature' in early-modern science that 'came to signal a granting of permission to enquiring minds' (pp. 56–7). In this passage, however, it appears that the link between knowledge, secrets and God is ultimately maintained.
45. Cavendish ([1666] 1991), p. 260.
46. Ibid., p. 288.
47. Ibid., pp. 258 and 262.
48. See Filmer, R. (1680), *Patriarcha: or the natural power of kings*, which, although not published until later, was widely circulated during the Civil War.
49. Cavendish ([1666] 1991), p. 260.
50. Ibid., pp. 269, 289 and 325–6.
51. Rogers (1996), pp. 199, 197 and 201. Rogers goes on to argue that Cavendish's science provides 'a utopia of female rule' in which feminised, internal rational matter is granted superiority over masculinised external brute force (pp. 201–3). However, he also suggests that *The Blazing World* provides 'An acknowledgement of the limitations besetting the separate sphere of female sovereignty', for while the Empress has command of her private world, 'The actual public world … proves much less tractable' (p. 207).
52. Cavendish ([1666] 1991), pp. 262–3.
53. Ibid., p. 263.
54. Gallagher, C. (1988), 'Embracing the absolute: the politics of the female subject in seventeenth-century England', *Genders*, **1**, Spring, 24–39, 27–8.
55. Cavendish ([1666] 1991), p. 263.
56. Ibid., 1991, pp. 288–9.
57. Ibid., 1991, pp. 295–6.
58. Milton, J. ([1667] 1971), *Paradise Lost*, ed. A. Fowler, London: Longman, **I**, l. 26, p. 44.
59. Ibid., **IX**, 1. 921, p. 492.
60. Cavendish ([1666] 1991), pp. 300, 336 and 339.
61. Ibid., p. 304. Cabalistic ideas were of central interest to a number of Cavendish's contemporaries, including John Dee (1527–1608) and Henry More (1614–87), both of whom are mentioned in this section (pp. 292–3 and 306), and Anne Conway (1631–79). In *New Atlantis* the laws of Bensalem are claimed to have been ordained by 'a secret cabala' (Bacon, 1857–74, vol. 3, p. 151).
62. Cavendish ([1666] 1991), p. 306.
63. Ibid., p. 308; compare with K. Philips's poetry about platonic female friendship. See also More, H. ([1659] 1987), *The Immortality of the Soul*, ed. A.J. Dordrecht, Boston and Lancaster: Martinus Nijhoff; who argues that souls may take voyages in and out of the body but, unlike *The Blazing World*, suggests that 'no man can when he pleases pass out of his Body thus, by the Imperium of his will' (p. 168).
64. Cavendish ([1666] 1991), pp. 306, 308–9.
65. Ibid., pp. 313 and 309.
66. Ibid., pp. 310–12.
67. Descartes, R. (1968), *Discourse on Method and the Meditations*, trans. and ed. F.E. Sutcliffe, Harmondsworth: Penguin, p. 112.
68. See Cavendish's *Grounds* (1668), which argues that 'the Rational Parts … must of necessity have the Properties and Nature of a Body' and that Man 'can have but

a parted knowledge, and a parted perception of himself' (pp. 239 and 55). Cavendish also troubles Cartesian epistemology in her poems (see Price, 1996, pp. 121–4). In *The Blazing World*, however, her attack is satiric and much more direct.

69. Descartes (1968), p. 95.
70. Cavendish ([1666] 1991), p. 313.
71. Ibid., p. 313.
72. Descartes (1968), pp. 164 and 112.
73. Cavendish ([1666] 1991), p. 313.
74. Ibid., p. 311. Compare with *Grounds* (1668), where Cavendish refers to the possibility of making, dissolving and governing worlds in the mind, identifying such worlds as 'Poetical Fancies', pp. 74–5.
75. Cavendish ([1666] 1991), p. 313.
76. Ibid., p. 348.
77. Ibid., p. 347.
78. Ibid., pp. 252–3.
79. Ibid., p. 348. Maston, J. (1997), *Textual Intercourse: Collaboration, Authorship and Sexualities in Renaissance Drama*, Cambridge: Cambridge University Press, also notes Cavendish's use of an apologetic mode when she tentatively affirms a 'self-sufficient' author position in the 'elaborate apparatus of self-signed letters' that introduce her *Plays* (p. 157). He argues that not only does Cavendish's use of the terms 'Authoress' and 'Poetress' 'to describe her vocation' contain a 'radical charge' given the sexual politics of authorship of the period, but that her striking emphasis on originality and 'home-grown genius' also 'draws on emergent paradigms of authorship', marking a break from, and implicit critique of, those which dominated the seventeenth century (p. 162).
80. Cavendish ([1666] 1991), pp. 347 and 252.

CHAPTER NINE

Gender, Science and Midwifery: Jane Sharp, *The Midwives Book* (1671)

Elaine Hobby

The early-modern midwifery manual is a fascinating phenomenon. It was from the first a cross-cultural affair, with key examples, such as Eucharius Rösslin's *Der Swangern frawen und he bammen roszgarten* ('the rose garden for pregnant women and midwives', 1513), and François Mauriceau's *Des Maladies des femmes grosses* (1668; translated as *The Diseases of Women with Child*), circulating widely not only in Latin, but in various European languages.[1] Though these books based many of their recommendations on ancient Greek medical models, they also rapidly incorporated findings from the new anatomy. Issued in a cheap octavo format, and in vernacular translations, they were an accessible source of information and advice on a wide range of matters concerning sexual life, including, for instance, methods to promote fertility, and also guidance on the pre- and post-natal care of the mother and child, as well as guidance on pregnancy and childbirth. The scale of interest in such books, and the breadth of their potential market, is indicated by the fact that Rösslin's *Roszgarten*, which first appeared in German, was, within 30 years, translated into Latin, Dutch, French, Spanish, Danish, Czech and, finally, English. Its English version, *The Byrth of Mankynd*, is extant in at least 11 editions dating from between 1540 and 1654, and many more variants have probably been lost.[2] Through midwifery manuals, understandings from both old and new science could reach a wide audience, and their prefaces and dedicatory epistles frequently urged readers not to make improper or indecent use of the materials they contained.

Apart from a section of the compilation *The Compleat Midwifes Practice* (1656), these books are the work of men.[3] Most of them were more or less unoriginal, regularly making use of illustrations copied from the much more expensive new anatomies, and constructing their written texts through a series of borrowings and paraphrases from those and other sources.[4] When Jane Sharp, claiming to be a midwife of 30 years' experience, brought out *The Midwives Book* in 1671 and became the first Englishwoman to publish a midwifery manual, she also followed such practices. In particular, she made use of Nicholas Culpeper's *A Directory for Midwives* (1651), and of his translations of Daniel Sennert's *Practical Physick* (1664) and Thomas Bartholin's *Bartholinus Anatomy* (1668), thereby producing a comprehensive account of conception, pregnancy, birth and the care of the newborn infant, and weaving into it instructions on how to make a huge variety of medical remedies, while peppering it with provocative anecdotes and asides.[5] Precisely because of *The Midwives Book*'s close relationship with its male-authored sources, it provides the modern reader with the opportunity not only to encounter a wide range of early-modern thinking about the body, but also to discover that the gender biases of scientific concepts were visible, to some extent, to Sharp and her contemporaries.

Before looking more closely at the ways in which *The Midwives Book* provides implicit and explicit commentary on the presence of masculinist bias in medical thought, it is necessary to sketch in a little more detail the book's relevant contexts. For, in publishing *The Midwives Book* in 1671, Sharp entered a medical world in flux. Patchily, traditional understandings of how the body worked – and so of how human reproduction could happen – were being challenged.[6] Although humoral explanations of how health was to be achieved and maintained continued to be fundamental to the logics of most medical practices, findings from the new anatomy presented other, competing models. From a modern perspective, midwifery writings from the period can therefore appear contradictory. Jakob Rueff, for instance, both follows the new anatomy in his assertion that the womb is held tightly in place by its ligaments ('horns'),[7] and draws on ancient Greek medicine in his view that uterine prolapse can be caused 'through wind inclosed in the Matrix, or corrupt humours'.[8] *The Compleat Midwifes Practice*, in its 1663 edition, laments the fact that other manuals depend on traditional beliefs instead of using knowledge of 'the Anatomical parts of the Body',[9] but also assumes that a woman's fertility is determined by her complexion, or humoral balance.[10] From the viewpoint of early-modern Europe, however, it must have seemed that the fundamental premises of such ideas were inherently compatible, and that they would, in due course, be found to be connected to each other.[11]

Alongside both of these ways of thinking were long-standing magical beliefs (some of which were also held by ancient Greek writers), for instance the idea that a pregnant woman's imagination had such powerful effects on her child's appearance that a baby could look like a man other than his father, if the mother was thinking of someone else at the time of conception.[12] It is such magical beliefs that underlie a variety of remedies in midwifery manuals, such as the common recommendation that an eaglestone (a stone within a stone) can be worn in pregnancy on the upper part of the body to prevent abortion, and then in labour attached to the lower body so as to attract the baby down the birth canal and hasten its birth.[13] Anne Finch, Viscountess

Conway, is today famous for her scientific and philosophical correspondence with Henry More. In 1658, when she was pregnant, her husband put considerable energy into obtaining an eaglestone for her, to reduce the risk of her miscarrying. Within days, he was reporting in a letter to his brother-in-law, Major Rawdon, that he had borrowed one.[14] Neither of the correspondents, nor Lady Conway herself, seemed to find any problem reconciling a confidence in the eaglestone's powers with her interest in the new science. Their concern, rather, was that news of the borrowed eaglestone might leak out, and that people would come to know of the pregnancy before a happy outcome could be trusted. The most famous example, of course, of the compatibility of what today seem contradictory models of the body's workings is the fact that William Harvey, who first established the circulation of blood, nonetheless continued to accept as well that the heart's motion was caused by its inherent pulsative faculty, a power it derived from the soul.[15] In the early-modern period, old and new versions of how the body worked were used together, and modern assumptions that a mechanical model was a necessary consequence of developments in scientific thinking were by no means yet established.

These changes in the medical model were also not the only important way in which the world of the early-modern midwifery manual was in flux. In the aftermath of the English revolution, where matters of religious belief both coded, and produced differences in, social and political attitudes, the extent to which divine intervention was assumed to be intrinsic to the body's well-being was also open to debate. In midwifery, the question of God's role was particularly pertinent: not only were midwives supposed to be licensed by the bishops (though it is probable that the majority of midwives sidestepped this requirement, and practised unlicensed), but tasks routinely expected of them included the performance of emergency baptisms, and the accompanying of new mothers when they went to be churched.[16] The midwife's job therefore combined the physical and spiritual care of both mother and newborn child, and in *The Midwives Book*, Sharp insists that a midwife be God-fearing as well as experienced: knowledge of the body's workings is necessary, but is not in itself sufficient.[17] Consistent with this perception of the body's spiritual dimension is the fact that, at the onset of labour, practical preparations for the baby's arrival are to take place, 'having first invoked the Divine assistance by whom we live and move and have our being'.[18] Although non-theocentric ways of understanding human existence were appearing, midwifery manuals were more likely than not to conceptualise birth in theological as well as medical terms.

If the emerging new anatomy, and shifting understandings of the mind–body relationship, created two mutating contexts for developments in the midwifery manual, there is also a further way in which the world entered by *The Midwives Book* was changing. Assumptions about the likely or necessary genders both of midwives and of authors were beginning to shift. In the earlier decades of the early-modern period, midwives were usually women, and midwifery manuals were normally written by men. By the 1670s, however, increasing numbers of men-midwives were appearing, beginning the trend that was to result, for instance in *Tristram Shandy*, in the stereotypes of the old widow midwife and the man-midwife with his bag of scientific instruments, and in an acceptance that the best attendants at a birth were male.[19] On the other hand, at the same time as men were entering the traditionally female sphere of the birthing chamber, increasing numbers of women were making inroads into the mainly male territory of the printed word.[20] The same year that *The Midwives Book* appeared, 1671, saw the publication not only of Aphra Behn's first two plays, but also the first edition of the best-selling *The Life and Death of that Excellent Minister of Christ, Joseph Alleine*, much of which was written by his widow, Theodosia Alleine. A Quaker, Elizabeth Hincks, also published her book of verse, *The Widows Mite*. Simultaneously, female-authored books already in print were being reprinted: 1671 saw the appearance of fine second editions of two early works by Margaret Cavendish, Duchess of Newcastle, *Natures Picture* and *The Worlds Olio*; a second edition of Elizabeth White's Nonconformist autobiographical narrative, *The Experiences of God's Gracious Dealings*; and the reissue of a work by the popular writer on cookery and household management, Hannah Wolley. It is not surprising, then, that whereas Jane Sharp's book is, to the modern eye, as contradictory as those of her contemporaries when it comes to the merging of humoral theory, new anatomy and magical beliefs, it evidences a keen awareness of gender politics, and implicit and explicit engagement with the position of women in the emergent new medicine.[21]

Indeed, from its opening pages, *The Midwives Book* takes an overt position on the question of the increasing appearance of men in the birthing chamber. The holy Scriptures, Sharp insists, citing Exodus 1, have 'recorded Midwives to the perpetual honour of the female Sex'.[22] Practising the art of midwifery is therefore 'the natural propriety of women', and the very idea that there could be men-midwives is rejected.[23] Where many male-authored midwifery manuals castigate midwives as '*MOTHER CARELESS*',[24] or insist that most of them are 'ignorant', and are responsible for complications that occur during birth,[25] Jane Sharp's midwife performs her work 'gently and prudently'.[26] Sharp also assumes, despite the apparently increasing practice of summoning male help if a labour became obstructed or prolonged, that it is the midwife herself who will deal with such difficulties, and that she will therefore develop skills in the use of surgical instruments.[27] *The Compleat Midwives Practice*, by contrast, has the midwife become a surgeon's assistant on these occasions,[28] and William Sermon deliberately omits any guidance on such matters in *The Ladies Companion* (1671), insisting that it is beyond a midwife's brief.[29]

A number of other differences between *The Midwives Book* and the male-authored tradition that it draws on also indicate a general awareness on Sharp's part of the implicit prejudices of received medical ideas about women. She firmly refuses, for instance, the performing of a caesarian section unless the mother has died, believing, quite rightly, that it would kill her.[30] Sennert, by contrast, leaves open to debate the question of whether to try to save the baby in this way,[31] and although Jacques Guillemeau also advises against caesarian section, he in addition informs the reader that he had tried the operation several times on live women before coming to the conclusion that it would always prove fatal.[32] Consistent with Sharp's position here on the importance of the mother's well-being is the fact that she also rejects the assumption that parents would prefer to have male rather than female children, remarking that mothers like to have daughters.[33] *The Compleat Midwife's Practice*, by contrast, seriously advises that if people want their children to be wise, the safest way to ensure this is to conceive only sons.[34] A further example of the way in which Sharp's medical judgements are informed by a wider gender politics occurs in her discussion of whether a mother should breastfeed her own children, or send them out to a wet-nurse. In male-authored texts, this is presented as a matter of ethics – for Sennert, for instance, maternal breastfeeding is a divinely-ordained duty[35] – but Sharp's approach is pragmatic. The mother's milk is likely, according to humoral theory, to suit her own child best, she explains,[36] but she also remarks: 'I cannot think it alwayes necessary for the mother to give her own Child suck; she may have sore breasts, and many infirmities, that she cannot do it.'[37]

A particular, and entertaining, difference of view is also found in accounts of a specific duty of the midwife, the cutting of the navel string (umbilical cord). It was traditionally believed that the length of cord left on the baby would determine the proportion of its sexual parts. In the male imagination, the midwife's resultant power was perhaps found frightening – *The Compleat Midwifes Practice* exhorting her not to cut the scythe too short,[38] and Culpeper advising those women who have exceptionally painful labours to blame the midwife who delivered them when they were babies themselves, for having left a short navel string, thus making their sexual parts too tight.[39] Sharp's perspective is much more practical, insisting that difficulties caused by the cutting of the cord are surely rare, since 'she must be a very unskilful Midwife that knows not how to tie and cut the Navel string'.[40] The severing of the cord is a simple matter: 'It is no matter what you cut it off with, so it be sharp to do it neatly.'[41] There seems, nonetheless, to be an element of glee in her specific recommendations concerning the connection between the cord's length and the size of the male genitals, that the cut should be made whilst 'alwayes remembring that moderation is best, that it may not be left too long, which may be as bad as too short'.[42]

The Midwives Book's refusal of some key conventions of medical thought is also found in its specifically literary characteristics. At the end of its first section, Sharp makes a promise that her book indeed delivers on: she omits 'hard names', that is, most of the Latin and Greek terms usually found in midwifery manuals, and writes 'as briefly and plainly as I can'.[43] She also draws attention to the fact that, in their anatomical sections, midwifery manuals regularly put descriptions of men's genitalia before those of women, and that in doing so they implicitly give greater importance to the male. She wryly remarks, 'perhaps when men have need of us they will yield priority to us'.[44] In her own book, she says, she will follow the traditional sequence, but only because she wants her work to be accessible even to those of the 'meanest capacity': that is, by implication, to the kind of stupid people who unthinkingly put men first.[45] It is consistent with this that when she comes to explain the Latin word for 'The Monthly courses of women', '*Menstrua*', she alludes to and refutes the standard false etymology in which 'menstrua' is connected to the word for monster, 'monstrum'. Women are not monstrous, Sharp insists; on the contrary, 'it is a Monstrous thing, that no creature but a women [*sic*] have them'.[46]

It is indeed in her discussions of the sexual parts and capacities of men and women that Sharp's book provides its clearest critique of both established and emergent medical ideas about the sexes. In many male-authored books, the penis (yard) is wondrous, being characterised by 'exquisite sense', and so capable of 'exquisite feeling'.[47] According to *Bartholinus Anatomy*, and as ancient Greek tradition also affirms, the testicles (stones) 'give strength and courage to Mens bodies',[48] and the penis is made from a substance that is 'peculiar and proper to itself'.[49] Sharp agrees that the male organ is 'peculiar', but she prompts the reader to interpret the word as meaning 'strange' rather than 'special'.[50] The yard she anatomises is a sickly and unpredictable organ, prone both to priapism and to other mishaps;[51] it is a part which 'swells with a windy spirit only, and riseth sometimes to small purpose'.[52] The medical bases of these readings are the same as those supporting men's books: in humoral theory, it is indeed 'windy meats', or flatulence-inducing foods, that are likely to stimulate erection, and the muscles of the penis tend to be described in a mechanical way, the '*Erector Penis*' contracting to effect a 'standing of the Yard'.[53] However, what Sharp makes of these medical 'facts' indicates that a very different interpretive model is at work from that used by her male predecessors and contemporaries.

Indeed, what is especially shown in *The Midwives Book* is a mischievous sense of irony at the importance credited to the penis in the masculinist model. Where Jane Sharp is of the view that, when it comes to the size of a penis, 'moderation is best', books by men tend to aspire to an organ of impressive length: nine inches in *The Compleat Midwifes Practice*, and eight or nine inches in Thomas Vicary's surgeons' manual.[54] One of Sharp's wittiest anecdotes tells how 'a *French* man complain[ed] sadly' that he was receiving less pleasure from sexual intercourse than he used to because, according to him, his wife's vagina had 'grown as a Sack'.[55] *The Midwives Book* goes on to reflect: 'Perhaps the fault was not the womans but his own, his weapon shrunk and was grown too little for the scabbard'.

It is a similar drive to cut male pretension down to size that causes Sharp to issue a warning which makes new meaning of the threatening 'hungry womb' that supposedly characterised women, whose sexuality was routinely represented as excessive or threatening.[56] *Bartholinus Anatomy* offers a particularly clear version of such beliefs:

> another action of the womb is said to be a certain Natural motion: whence *Plato* would have the womb to be a certain *Animal* or Live-wight, and *Aretius* saies it is an Animal in an Animal, because of its motion. For in carnal Copulation, and when it is possessed with a desire to conceive, it is moved now up and then down, and gapes to receive the Yard, as a Beast gapes for its Food.[57]

Where in male-authored medical writings such supposed uterine movements are fundamental to the ways women are perceived, Sharp turns this idea in a quite different direction. The head of the penis is the source of men's most intense sexual pleasure, she explains. During sex, she counsels, 'Let men be careful then how they enter too far, for it will be hard to say which were the greater loss, of the Stones [testicles] or the Nut [glans]'.[58] In her threat to the inconsiderate male reader that he might lose his sign of manhood if he enter too far, Sharp is turning such anti-women beliefs to produce instead the possibility of female retaliation.

It is not only in her use of the idea that the womb is a hungry animal that Sharp remakes the medical facts of women's bodies. Another common metaphor applied to the uterus in male-authored medical texts is to see it (following Aristotle) as a field into which 'seed' falls so as to produce conception. The Theodore Mayerne section of *The Compleat Midwifes Practice* for instance explains:

> We may observe the Natural Procreation of man to be altogether such, as we perceive the generation and beginning of Plants or Herbs of every kind to be: For as they, every one of them from the Seed of his kind, cast into the Womb of the Earth, do bud or increase, and do naturally grow to the perfect Form of this proper Nature: So man also being a reasonable Creature according to the quality of his body, doth naturally draw his Original and Beginning, from the sperm and seed of Man, projected and cast forth into the Womb of Woman, as into a field.[59]

Such an entirely passive presentation of a woman's role in procreation is not, of course, a necessary product of the medical model itself: not only was the uterus usually seen as active in its embracing of seed, but the woman, too, had to orgasm and thereby produce seed for conception to occur. In Sharp's account this different interpretation of the medical case is clear. She explains:

> The Womb is that Field of Nature into which the Seed of man and woman is cast, and it hath also an attractive faculty to draw in a magnetique quality, as the Loadstone draweth Iron, or Fire the light of the Candle, and to this seed runs the Womans blood also, to beget, nourish, encrease and preserve the Infant till it is time for it to be born.[60]

By this account, indeed, the woman, far from being a mere recipient for 'the sperm and seed of Man', plays a greater part than does the father in the child's formation: not only does she produce 'Seed also as well as the Man to sow the ground with',[61] she also provides the blood that nourishes the growing child. For *The Compleat Midwifes Practice* in 1656, by contrast, it is the man's seed that is 'the effective original of the creature'.[62] The anonymous male authors insist:

> In this place arises a question not trivial; whether the seed of woman be the efficient, or the material cause of generation? to which it is answered, that though it have a power of acting, yet that it receives the perfection of that power from the seed of man.[63]

The role of the woman's seed in this medical account is like that of Eve in the Bible (Genesis 2:18): it is a helpmeet to the male, giving 'a mutual assistance to the seed of man, in the work of generation'.[64]

Many other elements of Sharp's description of the female sexual parts work as an implicit refusal of the masculinist bias of established interpretations of medical ideas, ideas which in other respects she accepts. Her description of the female genitalia, for instance, reworks a bald mapping that she has found in Helkiah Crooke's *Microcosmographia*, producing in its place a celebration both of her sexual pleasure and of the accuracy with which a woman can urinate, thanks to her 'wings' or 'nymphs' (labia minora).[65] Whereas, for Crooke, a woman's urine 'runneth foorth in a broad stream with a hissing noise, not wetting the wings of the lap in the passage', Sharp provides a laudatory rewriting. In *The Midwives Book*, the rapid, graceful movement of the line imitates the movement of the urine, as it 'runs forth in a broad stream and a hissing noise, not so much as wetting the wings of the Lap as it goes along'.[66] In *Bartholinus Anatomy* this poetic rewriting of Crooke's account is given a different focus, urination being assessed, as it is in little boys' games, by measuring the distance achieved. The nymphs 'have charge of the Waters and Humors issuing forth. For between them as it were two walls, the urin is cast out to a good distance with an hissing noise, without wetting the Lips of the Privity.'[67] There are similar changes of perspective involved in Sharp's account of other parts of the female sexual anatomy, where delight is expressed over the clitoris, wherein 'lies the chief pleasure of loves delight in Copulation'.[68] Daniel Sennert, by contrast, although making mention of the clitoris, gives greater emphasis to the ancient belief that 'The place from whence comes life, is also the breeder of the most deadly poison'.[69] William Sermon is likewise disinclined to wax poetical over the female body, seeing as its positive feature the pleasure it can provide to men. He describes, for instance, the presence in the vagina of 'four Caruncles, resembling the leaf of the Mirtle-tree', which 'by tickling the Yard of a man, causeth the greater pleasure'.[70] Even so, the woman is not presented in terms that are particularly inviting. Sermon goes on to remark that the walls of the vagina are 'inwardly membranous and wrinckled, like to the inner skin of the upper Jaw of a Cows mouth'.[71]

It was an accepted medical fact in the early-modern period that women were more prone to illness than men because of their colder humoral balance. Whereas for Bartholin this difference between the sexes means that 'the womb is the Mother of many Diseases',[72] Sharp urges a quite different conclusion. Before discussing specific female ailments and their treatments she avers:

> Whosoever rightly considers it will presently find, that the Female sex are subject to more diseases by odds than the Male kind are, and therefore it is reason that great care should be had for the cure of that sex that is the weaker and most subject to infirmities in some respects above the other.[73]

This disinclination to accept the logics of mainstream interpretations of scientific 'fact' is also found in her jokey impatience with the 'impertinent disputes' of male authorities.[74] In *The Midwives Book*, though medical opinion often has to be accepted as objectively true, Sharp is also quick to point out that in many areas, 'physicians are at a stand' over how the body works.[75] She is also less inclined than her main source, Culpeper's *Directory for Midwives*, to be convinced by the 'pretty ways' of the more magically inclined Antoine Mizauld,[76] or to take as truth, like Sennert does, accounts of the Countess of Henneberg having borne 365 children.[77]

Sharp dedicated her book to her 'Sisters', the midwives of England.[78] What she provides for them, and for her more general female reader, is both a fairly up-to-date education in medical science, and a clear demonstration that all such learning should be approached with scepticism. She for one was not fooled by scientific 'facts', and she did not want other women to be fooled, either. It is important that modern commentators on such works do not make the mistake of assuming that our forebears were not just as aware as modern theorists are of the prejudices and preconceptions of medical thought.

Notes

Research for this chapter was funded by a grant from the Wellcome Trust, for which grateful acknowledgement is made.

1. Rösslin, E. (1513), *Der Swangern frawen und he bammen roszgarten*, Worms; Mauriceau, F. (1668), *Des Maladies des femmes grosses it accoucheés avec la bonne et véritable méthode de les bien aider*, Paris: J. Hénault. Hugh Chamberlen translated Mauriceau's book in 1672 as *The Diseases of Women*, but some subsequent English editions were entitled *The Accomplisht Midwife. Des Maladies* also appeared in the 1680s in German, Dutch, Italian and Latin. For the spectacular success of Rösslin's book, see Arons, W. (ed. and trans.) (1994), *When Midwifery Became the Male Physician's Province: the Sixteenth Century Handbook 'The Rose Garden for Pregnant Women and Midwives', Newly Englished*, Jefferson, NC: McFarland, pp. 2–3.
2. This first appeared in 1540, in a translation by Richard Jonas. The expanded 1545 edition, attributed to Thomas Raynold (otherwise Raynald), continued in print with small variations to 1654. There are extant editions dated 1552, 1560, [1564?], 1565, [1598?], 1604, 1626, 1634 and 1654. See J.W. Ballantyne [1907?], *'The Byrth of Mankynde': Its Author, Editions, and Contents*', first pub. in *Journal of Obstetrics and Gynaecology of the British Empire*, London: Sherratt and Hughes; Power, D. (1927), 'The Birth of Mankind or the Woman's Book: a Bibliographical Study', *The Library*, **8** (1), 1–37.
3. The female-authored section of *The Compleat Midwifes Practice* is a translation of extracts from a book by the midwife to the French royal family, Louise Bourgeois, which first appeared in French as her *Observations diverses* (1619): T.C., I.D., M.S., T.B. (1656), *The Compleat Midwifes Practice*, London: Nathaniel Brooke. A second, expanded edition, *The Compleat Midwife's Practice Enlarged*, by R.C, I.D., M.S., T.B., was published by Brooke in 1659, and again, further enlarged but with the same title, in 1663. The 1656 edition has page numbers 1–148, and then restarts at page 1, the section of Louise Bourgeois's work appearing in this second run of numbers, pp. 95–125. The Bourgeois selection is from Louise Bourgeois (1617), *Observations diverses, sur la sterilité, perte de fruict, foecondité, accouchements, et maladies des femmes*, 2 vols, Paris: A. Saugrain, **2**, pp. 1–71 and 199–251. For Bourgeois' book, see Perkins, W. (1996), *Midwifery and Medicine in Early Modern France*, Exeter: Exeter University Press.
4. *The Byrth of Mankynd*, for example, took its illustrations from Thomas Geminus (1545), *Compendiosa totius Anatomie*, London: J. Herford; which had copied them from Andreas Vesalius. Parts of *The Byrth*'s text come, via Geminus, from Thomas Vicary (1577), *A profitable Treatise of the Anatomie of mans body*, London: Henry Bamforde. For the Geminus–Vesalius connection, see Lind, L.R. (1949), *The Epitome of Andreas Vesalius*, New York: Macmillan. I am currently working on a history of the early-modern midwifery manual in English, tracing these and other connections.
5. All references to Jane Sharp (1999) are to *The Midwives Book: Or the Whole Art of Midwifry Discovered*, ed. E. Hobby, New York and Oxford: Oxford University Press. The introduction and notes to that edition give examples of interconnections with Sharp's sources, the key ones being: Thomas Bartholin (1668), *Bartholinus Anatomy*, trans. N. Culpeper and A. Cole, London: printed by

John Streeter; Daniel Sennert (1664), *Practical Physick*, London: Peter Cole; Nicholas Culpeper (1651), *A Directory for Midwives*, London: printed by Peter Cole; and *The Compleat Midwife's Practice*. There is discussion of the relationship between Sharp's book and those of male writers in Erickson, R.A. (1982), '"The Books of Generation": Some Observations on the Style of the British Midwife Book, 1671–1764', in Boucé, P.-G. (ed.), *Sexuality in Eighteenth-Century Britain*, Manchester: Manchester University Press, and Totowa: Barnes and Noble, pp. 74–94; in Fissell, M. (1995), 'Gender and Generation: Representing Reproduction in Early Modern England', *Gender and History*, **7** (3), 433–56; and in Keller, E.F. (1995), 'Mrs Jane Sharp: Midwifery and the Critique of Medical Knowledge in Seventeenth-century England', *Women's Writing: the Elizabethan to Victorian Period*, **2** (2), 101–11. These all offer provocative suggestions, but are based on rather limited reading.
6. See Sawday, J. (1995), *The Body Emblazoned: Dissection and the Human Body in Renaissance Culture*, London and New York: Routledge.
7. Jakob Rueff (1637), *The Expert Midwife or an Excellent and most Necessary Treatise of the Generation and Birth of Man*, London: Thomas Alchorn, p. 52.
8. Ibid., p. 68.
9. R.C. (1663), sig. a5.
10. Ibid., p. 252.
11. The usefulness of Eccles, A. (1982), *Obstetrics and Gynaecology in Tudor and Stuart England*, London and Canberra: Croom Helm, is reduced by a tendency to assess manuals on the basis of how modern their perception of reproduction is. A good brief guide to the coexistence of 'contradictory' medical models in this period is given by Porter, R. (1987), *Disease, Medicine and Society in England 1550–1860*, London: Macmillan. One of the most influential books on the history of sexuality in recent times, Laqueur, T. (1990), *Making Sex: Body and Gender from the Greeks to Freud*, Cambridge, MA, and London: Harvard University Press, makes fundamental errors through its failure to realise that more than one medical model was in play. An excellent account of the resultant shortcomings in Laqueur's thesis can be found in Park, K. and Nye R.A. (1991), 'Destiny is Anatomy', *The New Republic*, 18 February, 53–7. Park and Nye also point out that early-modern accounts of the body do not use the 'one-sex' model that Laqueur believes he has found in them, but rather a system of argument by analogy.
12. Sharp (1999), pp. 94–5. A useful survey of the continuation of such ideas into the eighteenth century is Boucé, P.-G. (1987), 'Imagination, pregnant women, and monsters, in eighteenth-century England and France', in G.S. Rousseau and R. Porter (eds), *Sexual Underworlds of the Enlightenment*, Manchester: Manchester University Press, pp. 86–100.
13. Sharp (1999), pp. 140–41, 146; Culpeper (1651), pp. 151–2. For this belief system, see Thomas, K. (1973), *Religion and the Decline of Magic*, London and New York: Penguin, p. 224.
14. Nicolson, M.H. (ed.) (1930), *Conway Letters: the Correspondence of Anne, Viscountess Conway, Henry More, and their Friends, 1642–1684*, London: Oxford University Press, pp. 122, 153–5. The relevance of Lady Conway's gender to her scientific correspondence is discussed in Martensen, R. (1994), 'The Transformation of Eve: Women's Bodies, Medicine and Culture in Early Modern England', in R. Porter and M. Teich (eds), *Sexual Knowledge, Sexual Science*, Cambridge and New York: Cambridge University Press, pp. 107–33. See also Crawford, P. (1990), 'The construction and experience of maternity in

seventeenth-century England', in V. Fildes (ed.), *Women as Mothers in Pre-Industrial England: Essays in Memory of Dorothy McLaren*, London and New York: Routledge, pp. 3–38.

15. Harvey, W. (1990), *The Circulation of the Blood and Other Writings*, trans. K.J. Franklin, intro. A. Wear, London and Vermont: Everyman, pp. xvii–xxv.
16. Notable recent work on midwifery's history includes Marland, H. (ed.) (1993), *The Art of Midwifery: Early Modern Midwives in Europe*, London and New York: Methuen; Donnison, J. (1977), *Midwives and Medical Men: A History of Inter-Professional Rivalries and Women's Rights*, London: Heinemann; Donegan, J.B. (1978), *Women and Men Midwives: Medicine, Morality, and Misogyny in Early America*, Westport, CT, and London: Greenwood Press; Towler, J. and Bramall, J. (1986), *Midwives in History and Society*, London and Dover, NH: Croom Helm; Evenden-Nagy, D. (1991), 'Seventeenth Century London Midwives: Their Training, Licensing and Social Profile', unpublished PhD thesis, McMaster University.
17. Sharp (1999), p. 11.
18. Ibid., p. 145.
19. See Sterne, L. ([1760] 1983), *The Life and Opinions of Tristram Shandy, Gentleman*, ed. I. Campbell Ross, Oxford: Clarendon Press, pp. 11–12, 84–8, 122. The assumption that men's use of instruments such as the Chamberlen family's obstetrical forceps constituted progress is now widely questioned. See Dally, A. (1991), *Women Under the Knife: a History of Surgery*, London: Hutchinson; Hess, A.G. (1994), 'Community Case Studies of Midwives from England and New England, *c.* 1650–1720', unpublished PhD thesis, University of Cambridge.
20. I have discussed women's writings of 1671 in Hobby, E. (2001a), '"Delight in a singularity": Margaret Cavendish, Duchess of Newcastle, in 1671', in *In-between: Essays and Studies in Literary Criticism*, **9**, nos. 1 and 2, 2001, pp. 41–62.
21. Other Restoration women writers also tackled medical topics. See Mary Trye (1675), *Medicatrix, or the Female Physician*, London: Henry Broome and John Leete; Sarah Jinner's almanacs (1658, 1659, 1660, 1664); Hannah Wolley (1674), *A Supplement to the Queen-like Closet*, London: Richard Lownds; Margaret Cavendish, Duchess of Newcastle, *The World's Olio* (1655), London: J. Martin and J. Allestrye; (1671) London: A. Maxwell. For an overview of such works, see Hobby, E. (1988), *Virtue of Necessity: English Women's Writing 1649–1688*, London: Virago; Ann Arbor, MI: University of Michigan Press, pp. 165–89.
22. Sharp (1999), pp. 11–12.
23. Ibid., p. 12.
24. Culpeper (1651), pp. 105–6.
25. T.C. (1656), p. 140.
26. Sharp (1999), p. 156.
27. Ibid., pp. 148–59.
28. T.C. (1656), pp. 112–19.
29. William Sermon (1671), *The Ladies Companion, Or The English Midwife*, London: Edward Thomas, p. 141. In earlier midwifery manuals it is assumed that such surgery is the midwife's job, as is indeed the case in Sermon's major source, whose instructions he erases: Jacques Guillemeau (1612), *Childbirth Or, The Happy Deliverie of Women*, London: A. Hadfield. For Sermon, such work belongs to men. For a study of a related change in the gendering of participants at caesarian births, see Blumenfeld-Kosinski, R. (1990), *Not of Woman Born: Representations of Caesarean Birth in Medieval and Renaissance Culture*, Ithaca and London: Cornell University Press.

30. Sharp (1999), p. 151. In Britain, caesarian section invariably proved fatal to the mother until 1793. See Eccles (1982), p. 115.
31. Sennert (1664), pp. 182–4.
32. Guillemeau (1612), pp. 187–8. A brief overview of the debate concerning the operation's viability appears in Eccles (1982), pp. 113–15. Some male writers also opposed caesarian section on live women; see, for instance, Jean Riolan (1671), *A Sure Guide*, trans. N. Culpeper, 3rd edn, London: George Sawbridge, p. 89.
33. Sharp (1999), p. 103.
34. R.C. (1663), p. 256.
35. Sennert (1664), p. 226.
36. Sharp (1999), p. 270.
37. Ibid., p. 265. See Fildes, V. (1986), *Breasts, Bottles and Babies: A History of Infant Feeding*, Edinburgh: Edinburgh University Press; and Fildes, V. (1988), *Wet Nursing: A History from Antiquity to the Present*, Oxford: Basil Blackwell. Moral reflections on maternal breastfeeding are discussed in *Breasts*, pp. 98–100, and *Wet Nursing*, pp. 87–9.
38. T.C. (1656), p. 16.
39. Culpeper (1651), p. 175.
40. Sharp (1999), p. 124.
41. Ibid., p. 166.
42. Ibid., p. 25.
43. Ibid., p. 65.
44. Ibid., p. 13.
45. Ibid., p. 13.
46. Ibid., p. 215. Crawford, P. (1981), 'Attitudes to Menstruation in Seventeenth-Century England', *Past and Present*, **91**, 47–73, is useful here, though Crawford misreads Sharp, believing she accepts the menstrua–monstrum comparison.
47. T.C. (1656), p. 17; Sermon (1671), p. 192; Culpeper (1651), p. 25.
48. Bartholin (1668), p. 57
49. Ibid., p. 59.
50. Sharp (1999), p. 25. *OED* indicates that both 'special' (sense A.3) and 'strange' (sense A.4) are current meanings of 'peculiar' in the period.
51. Ibid., pp. 23, 27, 28, 30, 31.
52. Ibid., p. 23.
53. The supposed musculature of the penis is explained in William Molins (1648), *Muskotomia: or, the Anatomical Administration of all the Muscles*, London: Edward Husband, p. 31. The phrase 'standing of the Yard' appears in Sharp (1999), pp. 25, 30.
54. Vicary (1577), sig. N1^{v}.
55. Sharp (1999), p. 46.
56. The history of seeing women's sexuality as dangerously excessive is sketched in Tuana, N. (1993), *The Less Noble Sex: Scientific, Religious, and Philosophical Conceptions of Woman's Nature*, Bloomington: Indiana University Press.
57. Bartholin (1668), p. 70.
58. Sharp (1999), p. 31.
59. R.C. (1663), pp. 243–4.
60. Sharp (1999), p. 53.
61. Ibid., p. 32.
62. T.C. (1656), p. 10.
63. Ibid., pp. 35–6.

64. Ibid., p. 63.
65. I have discussed this relationship in Hobby, E. (2003), 'Yarhound, Horrion, and the horse-headed Tartar: editing Jane Sharp's *The Midwives Book* (1671)', in K. Binhammer and J. Wood (eds), *Women and Literary History: For There She Was*, Newark: Delaware University Press. This passage is from Crooke (1651), p. 176.
66. Sharp (1999), p. 39.
67. Bartholin (1668), p. 77.
68. Sharp (1999), p. 41.
69. Sennert (1664), p. 114.
70. Sermon (1671), p. 195.
71. Ibid., p. 196.
72. Bartholin (1668), p. 70.
73. Sharp (1999), pp. 190–91.
74. Ibid., p. 105.
75. Ibid., p. 188; see also pp. 97, 118.
76. Ibid., p. 147; contrast Culpeper (1651), p. 159.
77. Sharp (1999), pp. 58, 78; contrast Sennert (1664), pp. 142–3.
78. Sharp (1999), p. 5.

CHAPTER TEN

The Masculine Matrix: Male Births and the Scientific Imagination in Early-Modern England

Ruth Gilbert

The following discussion explores early-modern images of male creation which were figured as intellectual or imaginative masculine births. It suggests that a profound gender uncertainty was embedded in masculine figurations of creation and procreation. The essay focuses in particular on the explicitly masculine tropes of vigour and virility that were routinely attached to the representation of the seventeenth-century new scientific project. It argues that this overt masculinisation of scientific discourse was, in fact, profoundly destabilised by its own enquiries into the creative process.

Early-modern men appropriated images of birth to describe both their textual production and their intellectual, technological and mystical aspirations and achievements. Writing, in particular, represented a form of production that negotiated male impulses towards creation without implicating them in the messier aspects of procreation. It could approximate a birth but ultimately circumvented the more grotesque characteristics associated with female parturition. For early-modern writers the equation between textual production and biological processes of reproduction was a standard literary convention. When, for example, Ben Jonson wrote his elegy to his dead son, the lost child was figured as his 'best piece of poetry'.[1] We might also think of Sidney's opening lines in 'Astrophil and Stella', in which the frustrated poet characterises himself as pregnant with poetry ('greate with childe to speake').[2] In such strategies the creative matrix is effectively transferred from the potentially appalling site of the female body (the womb) and becomes located instead in that part of the body which is seen to be the most masculine (the brain). The ideal male birth was thus clean, noiseless, bloodless and odourless. No afterbirth was expelled from a stretched and torn body. Instead a brainchild was produced, a perfect realisation of its masculine primogenitor.

However, early-modern representations of male births also suggest the limitations and complications of this model. Fantasies of masculine creation that tried to negate the female body, inevitably invoked the suppressed female presence. Figurations of male births repeatedly opposed the concept to the flesh, the ideal to the monstrous, the brain to the womb; but these distinctions could not be clearly marked or sustained. The reproductive man occupied a strangely paradoxical position. He was potentially a monstrous aberration, hermaphroditically effeminised by his engagement with the creative process. But he was also an absolute ideal: a superman whose hyper-masculinity was so impermeable that he could reproduce without the mediation of a female body.

This essay traces the instability inherent within such polarised representations. First, it discusses images of pregnant men as both a fantastic ideal of masculine creation and as its monstrous embodiment. In particular, it places such images within the context of early-modern scientific development. The discussion then moves on to a consideration of the roles of both Francis Bacon and William Harvey in the construction of the seventeenth-century new science. Both these figures, who have been associated with the most insistently masculine values of the new science, are shown to be rather more ambiguously placed in relation to the gendered discourse of that project. Through readings of Bacon's speculative fragment *Temporis Partus Masculus* and Harvey's *Exercitationes de generatione animalium*, I suggest that these explorations of the male role within the creative process in fact destabilise ideas of masculine scientific authority.

Both Bacon and Harvey played key roles in the development of an increasingly mechanised culture in the seventeenth century. Within such a context all origins were becoming uncertain. Questions about what it meant to be human, as well as what it meant to be male or female, were the focus of urgent inquiry. These themes are developed in the final section of this essay through an analysis of Milton's Satan who, in many respects, can be read as an archetypal new scientific hero pushing at the boundaries of knowledge and power. Drawing from this exploration of Satan as a consummate masculine creator, the essay concludes that his encounter with his grisly offspring, Sin and Death, is emblematic of the ways in which attempts to reify male creation in this period are necessarily mediated through, and disturbed by, images of embodied female reproduction.

Pregnant men: fantasies and horrors

Fantastical representations of autogenesis can be traced back to Lucian's *True History* (AD 2), in which he related that amongst the 'strange and wonderful things' that he had observed on his voyage to the moon was 'that fact that they [the moon people] are not born of women but of men'. Lucian's account included a surreal image of generation that anticipated Sir Thomas Browne's seventeenth-century expression of a desire to 'procreate like trees, without conjunction':[3]

> But I will tell you something else, still more wonderful. They have a kind of men whom they call the Arboreals, who are brought into the world as follows: Executing a man's right genital gland, they plant it in the ground. From it grows a very large tree of flesh, resembling the emblem of priapus: it has branches and leaves, and its fruit is acorns a cubit thick. When these ripen, they harvest them and shell out the men.[4]

The tree of flesh from which the Arboreals are grown is clearly, as Lucian notes, a priapic or phallic image. Men are born from the generative power of unmediated masculinity without the taint of passing through, or combining with, the female body. This idea, although expressed in the context of Lucian's satirical fantastic voyage, had a powerful hold on the imaginations of early-modern men. The phallic tree of life that generated new forms from the perfect male body was a recurring alchemical symbol, implying an absolute masculine creative purity. As late as 1676 Gabriel de Foigny repeated Lucian's image in his own imaginary voyage narrative, *Le terre Australe connue*, in which he described how the hermaphroditic people of his ideal society grew their children within them 'like Fruits upon Trees'.[5]

However, these idyllic forms of generation were the fantasies of distant and imaginary worlds. Images of male reproduction could equally suggest an unhealthy immoderation of the flesh. Underlying fantasies of male autogenesis was always the spectre of the monstrous birth. Ambroise Paré's popular collection, *On Monsters and Marvels* (1573), presented repeated stories and images of porous bodily forms. Here the tension between the classically sealed male body and the grotesquely open female body which, according to Bakhtin, is 'unfinished, outgrows itself, [and] transgresses its limits' was both described and displayed.[6] The paradox of male birth was encapsulated in the often reproduced contemporary image of a man with a head in his belly (Figure 10.1). This image of male pregnancy positions the brainchild within the male in an imitation of the unsealed, reproductive female body. But this figure, depicted in a stylised moment of revelation as the cloak is dramatically swept away from the body, bears little relation to the fleshy multiplication of either pregnancy or monstrous birth. The head stares out from the middle body as an addition, not a disruption, to the man who exhibits it. It is an emblem rather than an embodiment of male creation.

However, another image illustrated by Paré, the case of 'a man whose belly issued another man' (Figure 10.2), presents a far more grotesque image of male reproduction.[7] This figure was evidently based on a contemporary case. These conjoined twins were a famous Renaissance curiosity and toured

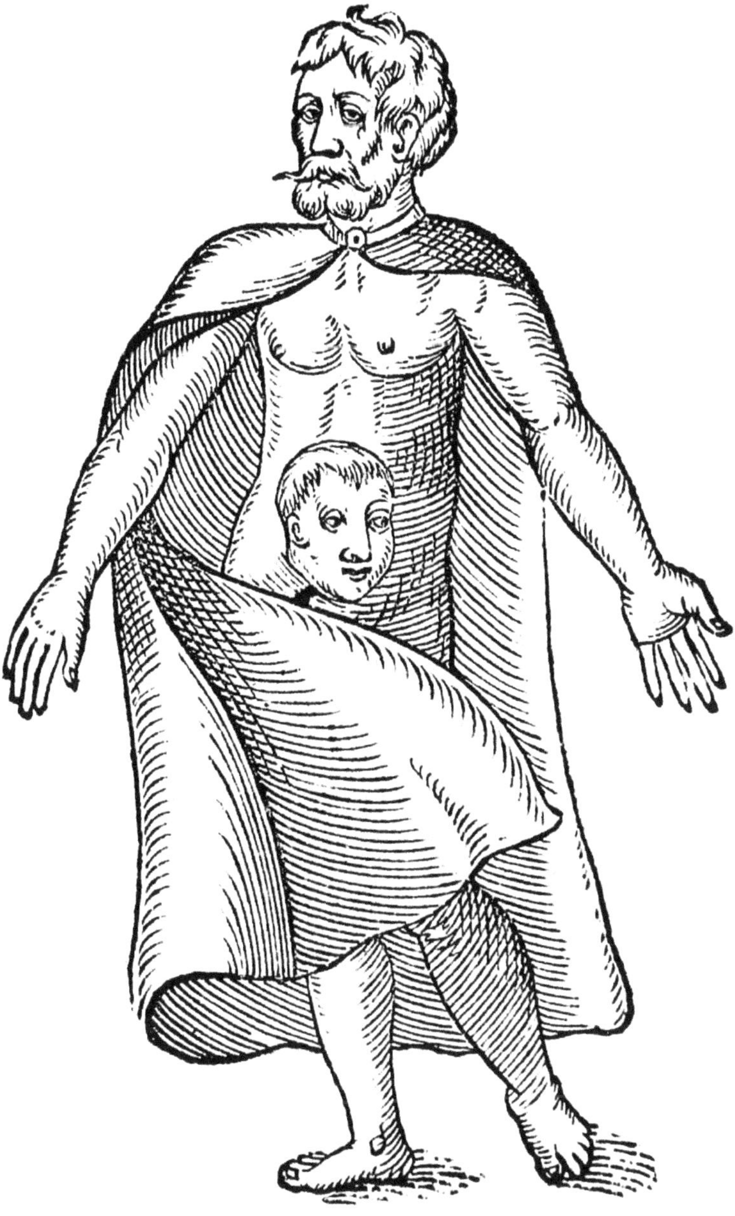

10.1 'Figure of a man having a head in the middle of his belly', in Ambroise Paré, 'Des Monstres et Prodiges', *Les Oeuvres* (Paris, 1579), vol. IX, p. cxxxi.

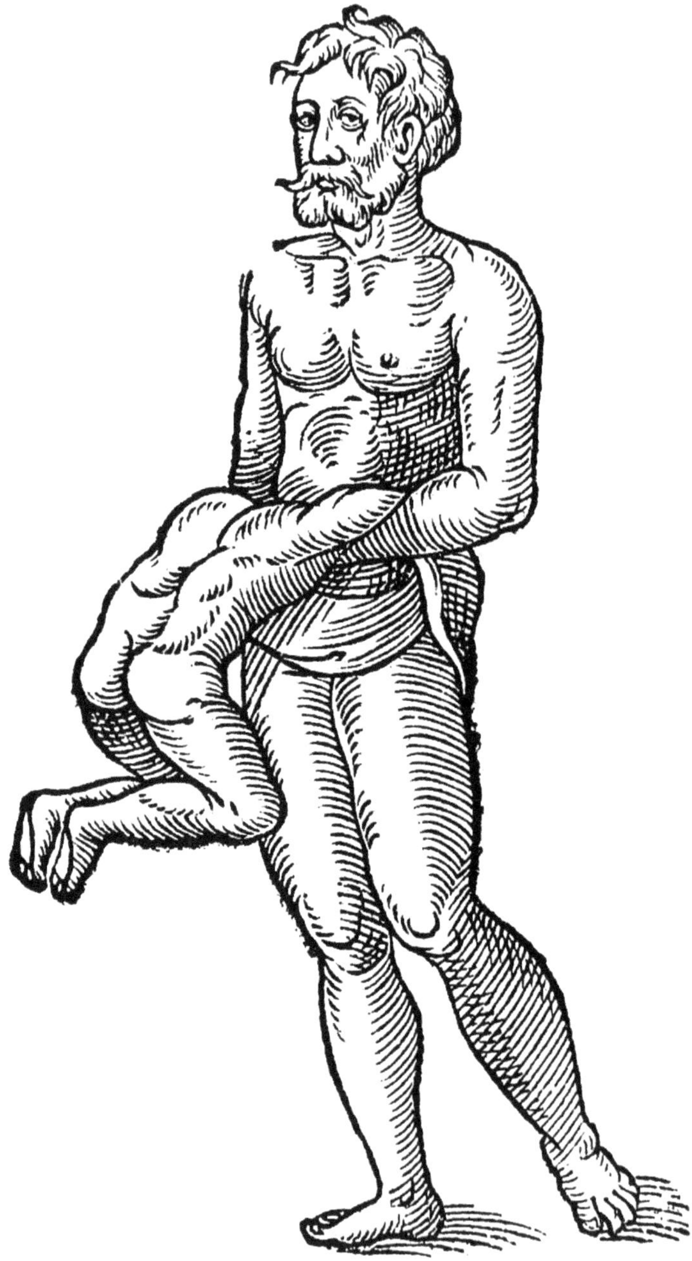

10.2 'Figure of a man from whose belly another man issued', in Ambroise Paré, 'Des Monstres et Prodiges', *Les Oeuvres* (Paris, 1579), vol. IX, p. cxxvi.

Europe extensively in the early seventeenth century. Martin Parker's illustrated broadside ballad 'The Two Inseparable Brothers' (1637) depicts the twins as fixed in a moment of ongoing birth, explaining that, 'the brother beares the brother'.[8] In his poem *Smectymnuus* (1662) John Cleveland similarly described this spectacle as 'Th' *Italian* Monster pregnant with his Brother'.[9] They symbolised a potentially unsettling, but clearly fascinating, image of reproductive excess and also implicitly recalled the female role in reproduction. According to the popular understanding of the time, they would have presented a tangible sign of maternal error: representing either the results of an overactive female imagination or a mismanaged excess of seed within the womb.[10]

These images of male pregnancy, whilst ostensibly escaping 'the trivial and vulgar way of coition' with women, which Browne had disparaged, were placed in a more complex relation to the grotesque female body.[11] By overspilling the boundaries of their own bodies and becoming associated with untidy reproduction, these men took on some of the grotesque characteristics of openness, excess and what has been termed 'leakiness' traditionally attributed to women in early-modern culture.[12]

Paré's images first circulated towards the end of the sixteenth century within what might be described as a pro-scientific culture. His collection of monsters and marvels drew from impressionistic, anecdotal and folklore discourses to discuss matters of medical and popular interest. By the end of the seventeenth century, as natural philosophy was increasingly framed by the epistemogical concerns of the new science, male pregnancy was reported less as a cause for wonder than as a matter of scientific curiosity and scrutiny. For example, in his *Treatise on Hermaphrodites* (1718), Giles Jacob included a letter written by Domat in 1697:

> I am at this very time employ'd in tending a Person of Quality that's come a great way off. In the right Side of his *Scrotum* he had a great Lump, bigger than the Head of a Child; which I cut off, and afterwards ty'd up the Spermatick Artery. This Lump was a Mass of Flesh, all over Spermatick, and very Solid, with very hard Bones in every part. 'Twas contain'd in an After-birth with a great deal of Water. The Spermatick Vessels which perform'd the Office of those we call Umbilical, were overgrown much beyond their natural size.[13]

He explained that the growth had developed when the man had been involved in a sexually active relationship with a woman but had not allowed himself to reach orgasm. The seed that should have been expelled had thus begun to develop into a baby within his own body. Domat's language of 'rational' scientific objectification describes the abjection of misplaced fertility resulting in the horror of self-impregnation. The lump was possibly a form of testicular cancer but in Domat's account, the male body had become, in the most grotesquely physical terms, an unnatural site of pregnant excess.

Subtending this disturbing image of (de)generation (which will lead to death not birth) was the misplaced female body, the proper receptacle of the sperm. Within the imperatives of enlightenment scientific rationalism the pregnant male body was no longer a cause for wonder, a sign of nature's infinite variety, or a symbol of masculine fecundity. It had been reframed as a medical phenomenon, a collection of parts to be catalogued and dissected. It was now a misfiring machine.

Concept and conception: Bacon, Harvey and the masculine birth of science

For the new science, masculine creation had to be figured as a product of intellectual rigour rather than as an encounter with monstrous embodiment. Domat's commentary typified how the scientific imperative was to become increasingly focused on reorganising Nature's idiosyncrasies into rational schemes. In order to achieve this aim the newly-born scientist had to be invested with heroic vision and courageous grandeur. Such status was inevitably associated with the assertion of extreme masculinism. Natural philosophy was held, as Cowley asserted in his ode 'To the Royal Society', to be a thoroughly 'Male-virtue'.[14] The establishment of the Royal Society in 1662, with its emphasis on Baconian scientific principles and objectivity, formed part of the political project to present the Restoration of Charles II as the resurrection of a supreme royal masculinity. Experimental science was consequently steered towards the ideology of the modern monarchy.[15] Science was, as Cowley depicted it, a hitherto neglected child, 'well bred and nurst but badly brought up on an effeminising diet of "wanton Wit", and "Desserts of poetry"'. In the logic of Cowley's ode, the virile paternal intervention of Bacon had saved the infant science for its realization in the vigorous work of the Royal Society.

Bacon was in many respects the obvious choice of primogenitor. His work presented an abundance of sexualised images which foreground a tradition that placed the potent masculine scientist in combat with a resisting female Nature.[16] However, the translated fragments of Bacon's *Temporis Partus Masculus*, or *The Masculine Birth of Time*, present a more complexly coded representation of gender and creation. This short experimental text takes the form of a prayer followed by philosophical musings from a father addressed towards a son. Bacon's fantasy is of the heroic birth of science:

> My dear, dear boy, what I plan for you is to unite you with *things* themselves in a chaste, holy, and legal wedlock. And from this association you will secure an increase beyond all the hopes and prayers of ordinary marriages, to wit, a blessed race of Heroes or Supermen.[17]

The male child which Cowley's ode later celebrated was anticipated in this dream of generative union between the scientist son and '*things* themselves'. According to Benjamin Farrington, 'the tacit insinuation [was] that older science represented a female offspring, passive, weak, expectant, but now a son was born, active, virile, generative'.[18]

However, the singular masculinity of the heroic male scientist is undoubtedly compromised in this image. As Evelyn Fox Keller argues: 'Behind the overt insistence of the virility and masculinity of the scientific mind lies a covert assumption and acknowledgement of the dialectical, even, hermaphroditic, nature of the "marriage between Mind and Nature".'[19] Within this 'marriage' the male scientist is feminised but never female; Bacon's creative matrix is located in the brain not the womb. Bacon's fatherly words, 'take heart, then, my son, and give yourself to me, so that I may restore you to yourself' construct a masculinist scientific register in which the seeds of wisdom are transferred in a dialectic exchange between passive and active subject positions.[20]

The mind, Bacon argues, is a palimpsest. It can only erase past errors by reinscription: 'On waxen tablets you cannot write anything new until you first rub out the old. With the mind it is not so: you cannot rub out the old till you have written in the new.'[21] Bacon represents the words of science as being imbued with a religious significance. Like Moses, the scientist will receive the truth passed on holy tablets from the father to the son, but only when this new truth has been received can all misconceived golden calves be destroyed. The fantasy of a receptive mind, purged of 'unholy' and 'unclean' errors and idols of the past is effected through an entirely discursive generation.[22]

As Thomas Sprat asserted in his *History of the Royal Society* (1667), by rigorously following Baconian principles of clarity and objectivity, science itself was to be heroically reborn. But, as Bacon had suggested, this epistemological rebirth was dependent on a fundamental process of preconception. The Royal Society, Sprat claimed, would banish error and restore truth: 'separat[ing] the knowledge of *Nature*, from the colours of *Rhetorick*, the device of *Fancy*, or the delightful deceit of *Fables* … They [the members of the Royal Society] have attempted to render it an Instrument, whereby Mankind may obtain a Dominion over *Things*.'[23] Sprat's zeal reflected the Royal Society's seemingly compulsive energy for collecting, classifying and organising the natural world. If dominion over things was to be fulfilled then all things had to be seen and known. Masculine science would look unremittingly into female Nature's unruly recesses and find new ways to recreate what it found.

William Harvey was given a key role in this new scientific programme of harnessing the natural world into the Restoration programme of regeneration. Harvey was acclaimed as a central figure in the construction of the new scientist as a masculine hero; he was cast as Bacon's heir, and was famously celebrated in Cowley's ode, 'Upon Dr. Harvey', as the heroic penetrator of 'Coy Nature'. His *Exercitationes de generatione animalium* (1651) presented an inquiry into what Harvey termed the 'dark, obscure business' of conception.[24] Harvey prefaced this text with the confident assertion that 'Natures Book' was 'open and legible', arguing that scientific scrutiny would allow entry into 'her Closet-secrets' (sig.A2^{v}). However, the secret that Harvey revealed considerably undermined such masculine self-assurance. Through a close examination of the embryology of chickens, Harvey's study proposed, albeit inconclusively, that life was produced epigenetically: that is, it developed within the female egg. This emphasis on the primacy of the material egg resulted in a description of generation in which men were almost incidental participants.

In his final chapter entitled 'Of the conception', Harvey puzzled at length over the male role within the creative process. Masculine participation in generation was figured as a disorientated journey within the female body. Harvey could not understand where the sperm went when it entered the dark recesses of woman:

> But I cannot but wonder, where *faculty* … when the act of *coition* is finished, before the production of the *Egge* or *Conception* doth reside? and to what that *active vertue* of the *male* is imparted? namely, whether to the *Uterus* alone or to the whole *Female*? or rather, primarily to the *Uterus*, but secondarily to the *female*? or lastly, whether, as we see with our *eyes*, and think with our *braines*, so a *female* doth conceive with her *uterus*?[25]

Harvey's difficulty in locating the sperm (the lost male presence) was resolved by placing the brain (gendered as male) as an original which was the model for uterine reproduction. He thus proposed that the brain was a generative matrix and presented an extended analogy between the brain and the womb.

Harvey's argument became increasingly abstract as he observes that both concept and conception are '*Immaterial*' created from '*phansie*':[26] the brain conceives in order to produce art, whilst the womb conceives in order to produce an embryo. His observations into the wombs of deer after conception, conducted in an experiment demonstrated to Charles I, had shown that no trace of seed could be observed within the uterus. He concluded, 'since I plainly see that nothing at all doth remaine in the *Uterus* after *coition*, whereunto I might ascribe the principle of *generation*; no more remaines in the *braine* after *sensation* and *experience*'.[27] The reinscription of masculinity into generation was thereby presented in an equation between brain and womb, concept and conception, which was predicated on absence. The heroic male scientist, unremittingly exploring Nature's darkest recesses, was forced to conclude his study by reflecting on that which could not be seen or known.

Harvey's text thus presented a somewhat uneasy, if unresolved, case for female primogenesis. This was a potentially untenable position within the explicitly masculine register of the new scientific project. However, *de generatione* presents a series of textual strategies which attempted to create a distance between the text and its uncomfortable conclusion. *De generatione* was framed textually and pictorially by representations of male births that reclaimed creation as masculine. The frontispiece image, which depicts Zeus opening the cosmic egg, summarises the tension between scientific heroism and the reproductive female body (Figure 10.3). The god sits high on a pillar holding the egg in which life is encapsulated. The egg is the feminine agent of birth but it is held and controlled by the supreme masculine generatrix. The creatures released from it (a reptile, snake, child, insect, bird and goat) encapsulate all forms of life. The writing on the shell, '*ex ova omnia*', proclaims Harvey's thesis that all life springs from the egg. But the masculine inscription on this symbol of female generation also reminds the reader who owns the concept if not the conception.

This insistence on the supremacy of masculine creation is further enforced before Harvey's thesis is presented. Martin Lluelyn's prefatory poem signals the childless Harvey's role as the consummate masculine scientist whose textual offspring confirms his generative powers:

> Live Modern Wonder, and be read alone,
> Thy *Braine* hath *Issue*, though thy Loins have none.
> Let fraile *Succession* be thy Vulgar Care;
> Great *Generation's* selfe is now thy Heire. (sig. A4)

The scientist is thus placed as the primogenitor in a tradition in which male textual production is privileged over female bodily reproduction. Harvey is celebrated as the singular father of all generation. Lluelyn hails him as the ultimate figure of male autogenesis: the man who can 'scape the *Woman*'. But Harvey's study had revealed that this was a troubled proposition. Beyond the tropes of poetry and the scientific imagination, what kind of brainchild could the masculine scientist legitimately create?

Man-made life: mechanisation and creative anxieties

The work of Bacon and Harvey articulated a more complicated gendering of scientific creation than has often been assumed. In spite of (or perhaps because of) these less confidently masculine sub-texts, one of the implicit aims of the seventeenth-century scientific programme was to reinstate the supremacy of male creation. Susan Bordo has explored the seventeenth-century development of a scientific culture in terms of a narrative of birth,

10.3 'Zeus opening cosmic egg', in William Harvey, *Exercitationes de Generatione Animalium* (London, 1651), frontispiece.

describing it as a 'drama of parturition' in which the maternal body became progressively estranged and disavowed, resulting in 'the Cartesian *re*-birthing and re-imagining of knowledge and the world as masculine'.[28] Whilst focusing on the gynophobia present within early-modern culture, Bordo also suggests that the masculinisation of knowledge was predicated on a desire for, as well as a disgust with, femininity. If female nature was to be effectively subdued by masculine science then creation had to be authoritively coded as male.

Within the new science, nature became viewed as matter which could be fashioned by masculine art, or *techné*. So the scientist 'Fathers of Salomon's House' in Bacon's *New Atlantis* (1627), for example, are celebrated for their deposition of the creative power of female nature. They alter and recreate the beasts, birds and plants that are held within their experimental domain.[29] As Sprat later suggested, 'dominion over things' was the proper aim of early-modern scientific endeavour. Within such a framework the role of women in creation became increasingly problematised. By proposing a masculine control over nature, the new science sustained ideas about the potential of exclusively male creation that echoed older, more mystical, ways of thinking. Jewish mysticism and alchemical experimentation, for example, both attempted to create life outside of the female body. Writings about both the golem (an animated creature formed from clay in the Jewish tradition) and the homunculus (the alchemical 'little man') represented fantasies of man-made life which recurred during the seventeenth century and beyond.

Throughout the seventeenth century anatomical, philosophical and poetic explorations of the human body figured it as a kind of machine. Experimental anatomy had opened the body up and revealed that it was, like the inside of a clock, a collection of interlocking parts. Harvey's *De Motu Cordis* (1628), which proposed that the blood circulated through the heart like a pump in a mechanistic process, marked a key moment in the contemporary reconceptualisaiton of the human condition. Such a profound shift in thinking about the body provoked equally profound questions about human origins. As Jonathan Sawday has put it:

> If God still existed, then he was certainly no architect, perhaps not even a creator in the older sense of a fashioning and forming deity. Rather he was a mechanic, an engineer, a watchmaker even, whose presence was no longer required for the continuing operation of the orderly movement of the machine.[30]

As creation came to be seen as an increasingly mechanistic process so the creator was implicitly a male scientist. As Elaine Hobby makes clear, in her essay in this volume (see Chapter 9), mechanistic models of the body in this period were neither fixed nor absolute and coexisted with other (sometimes seemingly contradictory) belief systems. Nevertheless, Bacon's old adversary, female Nature, had become for Boyle, by the late seventeenth century, 'God's great pregnant Automaton'.[31] But the representation of human life as an automatic process evoked potentially confusing and unsettling reactions. If the body was a machine then what constituted humanity?

Descartes attempted to resolve his own disturbing vision about mechanisation by proposing how the human body-machine could be distinguished from artificial life forms. His argument was that whilst animals ('these natural automata') could be artificially created, a mechanistic human, 'which had a likeness to our bodies and imitated our action', would be recognised as a false construction of a person because it would lack language and reason.[32] However, both these qualities – which were for him the only means of defining humanity – were also imperceptible, and Descartes' recurring anxiety was that he would misrecognise a machine as human. He expressed a paranoid fantasy in which the boundaries of humanity were dissolute and deceptive:

> If I chance to look out of a window on to men passing in the street, I do not fail to say, on seeing them, that I see men … and yet, what do I see from this window, other than hats and cloaks, which can cover ghosts or dummies who only move by means of springs?[33]

The fantasy of autogenerative creation was rearticulated by Descartes as a fear of self-deception. In a world in which the human body was no more than a collection of parts, which could (in theory) be imitated by man-made forms, the early-modern masculine creator sacrificed a secure faith in the authenticity of origins. The reiteration of stories about masculine creation had gone beyond a problematising of femininity or an appropriation of traditionally-female generative power to a point of exhaustion as all origins became insecure and questionable. The masculine matrix had recreated itself, in stories and in science, as a reproductive machine.

The 'All-One Paternall': returning to origins

In this context of increasing ontological and epistemological uncertainty, early-modern writers continued to return to Judaeo-Christian religious principles as a way of interpreting their shifting world. The primogenitor of all early-modern fictions of male birth was YAHWEH, the original 'artificier', patriarchal golem-maker and divine generative power. Du Bartas celebrated God's singular creation of Christ, his son, as a sublime idea, a perfect origin:

> For sans beginning seed, and Mother tender,
> This great Worlds Father he did first ingender,
> (To wit) his Sonne, Wisedome, and Word eternall,
> Equall in Essence to th'*All-One* paternall.[34]

Such fantasies of male creation were intrinsic to the Judaeo-Christian tradition. These were spiritual births of the masculine imagination.

When Eve is born in Milton's *Paradise Lost*, Adam is expectant, yielding and penetrated. In a state of suspended consciousness, Adam witnesses the birth as an inner vision: 'Mine eyes he clos'd, but open left the cell / Of fancy my internal sight', suggesting that this original male birth was created from within Adam's imagination (VIII.460–61).[35] The idea of woman was born of Adam and also within him. The moment of separation when another was created from the self was also the point from which men would be of woman born. The creation of the first mother thus also represented the last male birth. Fantasies of autogenesis rehearsed a fictional return to a pre-original condition in which this determining maternal body could be suppressed.

In *Paradise Lost* birth is authorship. God, 'thee author of all being' (III.374), ignites Milton's cycle of male birth producing both Eve's address to Adam as 'My author' (IV.635), and Sin's monstrous claim to Satan: 'Thou art my father, thou my author, thou / My being gavest me' (II.864–5). Troubled origins saturate the narratives of the poem, presenting a series of contradictory and contested versions of authorship. Satan's challenge to God's authority is posed as a question about ontological origins:

> Who saw
> When this creation was? Remember'st thou
> Thy making, while thee maker gave thee being?
> We know no time when we were not as now;
> Know none before us, self-begot, self-rais'd
> (V.856–60)

This demand for ocular proof of an external maker echoes the language of Baconian plunder and Harveian enquiry into the reproductive body, to be 'confer[ed] with our own eies'.[36] *Paradise Lost* was published in 1667, the same year as Sprat's *History of the Royal Society*, and Satan speaks with the voice of the new scientist. His 'unspeakable desire to see, and know' (III.662) leads him to explore and colonise the universe in a quest for knowledge and power.

But Satan's fantasy of autogenesis becomes hideously realised in his encounter with his monstrous progeny, Sin and Death. This degenerate male birth, the issue of incest and abjection, is placed in contrast to the later narrative of Adam giving birth to Eve. Paradigms of male creation which were based on fantasies of classical androgynous disembodiment are inverted when we meet Satan's creations: Sin and Death represent a grotesque expression of uncontrolled bodily proliferation.

It is no coincidence that Satan meets his offspring on his journey towards the confusion of boundaries which is Chaos: 'the womb of nature and perhaps her grave' (II.911). Sin embodies Julia Kristeva's description of the abject horror and fascination associated with female birth, which she terms 'a scorching moment of hesitation'.[37] This monstrous female is posted at the Gates of Hell, the precise limen between worlds. Within this context of collapsing and porous boundaries Satan's meeting of his grim anti-family is a telling misrecognition. Satan's story of autogeneration is no longer his own and it is his prodigious offspring who must relate this narrative of origins:

> Out of they head I spring: amazement seized
> All the host of heaven; back they recoiled afraid
> At first, and called me Sin, and for a Sign
> Portentous held me;
> (II.758–61)

So Satan's brainchild, born like Athena from Zeus, from the left side of her father's head, tells the story of her own birth.[38]

In contrast to Adam's painless generation of Eve as an abstract vision, Sin's fiery entrance into life is a confusing and painful parturition. As she tells her father:

> All on a sudden miserable pain
> Surprised thee, dim thine eyes, and dizzy swum
> In darkness, while thy head flames thick and fast
> Threw forth, till on the left side opening wide,
> Likest to thee in shape and countenance bright,
> Then shining heavenly fair, a goddess armed
> Out of thy head I spring
> (II.752–8)

The grotesque births which Sin's body continually bears are the results of Death's incessant and incestuous rapes. These creatures, which 'howle and gnaw' at their mother's bowels in a monstrous frenzy of destructive reproduction, are the most dramatically abject creations within the poem (II.799). The story that Sin embodies, and narrates, is of sexual degeneracy, incestuous union and monstrous creation. It is in every respect an inversion of dreams of pure male generation. Sin's hideous form and presentation of distorted and debased maternity clearly echo Spenser's serpentine female monster, Error.[39] Sin is Satan's own error passed on to humanity: she is the primary sign of a misconceived creation.

In representing these original moments of male reproduction Milton suggests the limits as well as the possibilities of masculine creation, procreation and (re)creation. Early-modern representations of male autogenesis could only ever flirt with the fantasy of erasing the female presence in reproduction. As *Paradise Lost* suggests, the procreative female body, 'this fair defect of Nature', could be denied and disavowed but not escaped. The masculine matrix was finally most fertile in the generation of narratives about its own reproduction. In 1671, four years after the publication of *Paradise Lost* and Sprat's *History of the Royal Society*, Jane Sharp published *The Midwives Book*. As Elaine Hobby demonstrates, Sharp's intervention forcibly reinstated the role of women at the centre of early-modern reproduction. Images of birth could be appropriated, but ultimately, in the postlapsarian world, children were always of women born.

Notes

1. Jonson, B. (1975), 'On My First Son', in *Poems*, ed. I. Donaldson, Oxford: Oxford University Press, pp. 26–7, l. 10.
2. Sidney, Sir P. (1962), 'Astrophil and Stella' in *Poems of Sir Philip Sidney*, ed. W.A. Ringler, Oxford: Clarendon Press, p. 165.
3. Browne, Sir T., *Religio Medici* (1977), in *Sir Thomas Browne: The Major Works*, ed. C.A. Patrides, Harmondsworth: Penguin, p. 148.
4. Lucian (1927), *A True Story (Verae Historiae)*, trans. A.M. Harmon, 8 vols, London: William Henemann, vol. 1, pp. 275–7.
5. Foigny, G. de [James Sadeur] (1693), *A New Discovery of Terra Incognita Australi, or the Southern World*, London: n.p., p. 85.
6. Bakhtin, M. (1968), *Rabelais and His World*, trans. H. Iswolsky, Cambridge, MA: MIT Press, p. 27.
7. Paré, A. ([1573] 1982), *On Monsters and Marvels*, trans. J.L. Pallister, Chicago: Chicago University Press, p.10.
8. Parker, M. ([1637] 1927), 'The Two Inseparable Brothers, or a Strange Description of a Gentleman Who hath an Imperfect Brother Growing out of His Side', in H.E. Rollins (ed.), *The Pack of Autolycus*, Cambridge, MA: Harvard University Press, pp. 7–14 (p. 13).
9. Cleveland, J. (1967), 'Smectymnuus, Or the Club-Divines', in *The Poems of John Cleveland*, ed. B. Morris and E. Withington, Oxford: Clarendon Press, pp. 23–6, l. 25.
10. For more on the power of the female imagination in the creation of monstrous births, see Huet, M.-H. (1993), *Monstrous Imagination*, Cambridge, MA: Harvard University Press. Paré places both figures in a chapter entitled 'An example of Too Great a Quantity of Seed'.
11. Browne (1977), p. 148.
12. For an extensive discussion of women as overspilling the boundaries of their bodies see Paster, G.K. (1987), 'Leaky Vessels: the Incontinent Women of City Comedy', *Renaissance Drama*, **18**, 43–65.
13. Jacob, G. (1718), *Tractatus de Hermaphroditis: Or, a Treatise of Hermaphrodites*, London: n.p., pp. 86–7.
14. Cowley, A. (1905), 'To the Royal Society', in *Abraham Cowley: Poems*, ed. A.R. Waller, Cambridge: Cambridge University Press, p. 448. Cowley's Ode was originally published with Thomas Sprat's *History of the Royal Society* (1667), London.
15. See Sawday, J. (1995), 'Autogenesis: the Masculine Discourse of Science and Reason', in J. Sawday, *The Body Emblazoned: Dissection and the Human Body in Renaissance Culture*, London: Routledge, pp. 230–48.
16. For an exploration of the gendering of science and nature see Merchant, C. (1980),

The Death of Nature: Women, Ecology and the Scientific Revolution, San Francisco: Harper & Row; and Keller, E.F. (1985), *Reflections on Gender and Science*, New Haven: Yale University Press.

17. Farrington, B. (1951), '*Temporis Partus Masculus:* An Untranslated Writing of Francis Bacon', *Centaurus*, **1**, 193–205 (p. 201).
18. Ibid., p. 194.
19. Keller, E.F. (1980), 'Baconian Science: A Hermaphroditic Birth', *The Philosophical Forum*, **3**, 299–308 (p. 305).
20. Farrington (1951), p. 201.
21. Idem.
22. Graham Hammill argues that Bacon's emphasis on purging the intellect in this text and others in the Baconian corpus also relates to 'a sexuality of purging, the experience of which creates an erotic link between men' (p. 238): Hammill, G. (1994), 'The epistemology of expurgation: Bacon and *The Masculine Birth of Time*', in J. Goldberg (ed.), *Queering the Renaissance*, Durham, NC and London: Duke University Press, pp. 236–52.
23. Sprat, T. (1667), *The History of the Royal Society of London, For the Improving of Natural Knowledge*, London, pp. 61–2.
24. Harvey, W. (1653), *Anatomical Exercitations, Concerning the Generation of Living Creatures*, London, p. 539.
25. Ibid., p. 540.
26. Ibid., pp. 543 and 545.
27. Ibid., p. 546.
28. Bordo, S (1986), 'The Cartesian Masculinization of Thought', *Signs: Journal of Women in Culture and Society*, **11**, 439–56 (p. 441).
29. Bacon, F. (1857–74), *New Atlantis*, in *The Works of Francis Bacon*, ed. J. Spedding et al., 14 vols, London: Longman, vol. 3, pp. 158–69.
30. Sawday (1995), p. 29.
31. Robert Boyle. Quoted from Bordo (1986), p. 453.
32. Descartes, R. ([1649] 1981), 'Letter to More, 5 February, 1649', in *Descartes: Philosophical Letters*, ed. and trans. A. Kenny, Oxford: Basil Blackwell, p. 244; Descartes, R. (1968), 'Discourse 5', in *Descartes: Discourse on Method and the Mediations*, ed. and trans. F.E. Sutcliffe, Harmondsworth: Penguin, pp. 73–4. See also Descartes's letter to Reneri for Pollut (1638), in *Philosophical Letters*, pp. 53–4.
33. Descartes, 'Second Meditation', in *Discourse on Method and the Mediations*, p. 110.
34. Du Bartas (1979), *The Divine weeks and Works of Guillaume de Saluste Sieur Du Bartas*, trans. J. Sylvester, ed. S. Snyder, 2 vols, Oxford: Clarendon Press, vol. 1, p. 114.
35. Milton, J. (1980), *The Complete Poems*, ed. G. Campbell, London and Melbourne: Dent, Everyman.
36. Harvey (1653), 'Preface', sig. A2v.
37. Kristeva, J. (1982), *Powers of Horror: An Essay on Abjection*, trans. L.S. Roudiez, New York: Columbia University Press, p. 155.
38. For more extensive discussions of the origins of Sin in *Paradise Lost*, see Chambers, A.B. (1963), '"Sin" and "Sign" in *Paradise Lost*', *Huntingdon Library Quarterly*, **26**, 381–2; and Gallagher, P.J. (1976), '"Real or Allegoric": the Ontology of Sin and Death in *Paradise Lost*', *English Literary Renaissance*, **6**, 317–35.
39. For descriptions of Error that compare with Sin, see Edmund Spenser, *The Faerie Queene*, I.i.14.

CHAPTER ELEVEN

From Nymph to Nymphomania: 'Linear Perspectives' on Female Sexuality

Bettina Mathes

De nymphomania is the title of Johann Nietner's dissertation, which he submitted to the University of Erfurt in 1694. Nietner's 64-page long (or short) treatise deserves attention because it is one of the first texts to introduce the term 'nymphomania' for a disease that since the times of Hippocrates had figured under the name 'hysteria' or 'furor uterinus'.[1] By the eighteenth century nymphomania had become well established as a synonym for 'furor uterinus' in most of the major medical lexicons, for example Johann Jakob Woydt's *Schatzkammer Medicinisch-und natürlicher Dinge* (first published in 1701) or Albrecht von Haller's *Medizinisches Lexicon* (1756). Both register the term 'nymphomania' but refer the reader to 'furor uterinus' for its description.[2] This practice suggests that the introduction of the term nymphomania does not indicate the discovery of a new disease, rather it reflects changes in the perception of the female genitalia – that is, the clitoris and labia (the so-called nymphae) – and helps to make visible a significant part of their history.

Nymphomania and nymphae

Nietner defines nymphomania as a 'delirium accompanied by an extraordinary desire for copulation'.[3] The symptoms range from sadness, restlessness and melancholy to delirious sexual agitation. The *Medizinisches Lexicon* describes nymphomania as an excessive form of compulsory heterosexuality: 'They [the nymphomaniacs] constantly fall in love with men; as soon as they see or think of them they lose their mind: ... They exhibit their bodies without any shame, they challenge men, run after them even on the streets, they attack them like wild beasts'.[4] As the *Schatzkammer* explains, nymphomania is most often caused by 'a lack of sexual intercourse; and this is why virgins and widows as well as those whose husbands are weak suffer from this disease'.[5] The heterosexual context of the disease is further reflected in its therapy. The *Schatzkammer* recommends sexual intercourse 'as the nicest and most successful therapy'.[6] For widows or religious women, who were supposed to live a celibate life, genital massages by a physician or midwife were recommended.[7] In nosology and therapy, nymphomania did not differ from hysteria. What was new here about nymphomania was its etymology. In contrast to hysteria or furor uterinus – terms that refer to the uterus – nymphomania draws attention to women's external genitals. According to Nietner, nymphomania goes back to 'the Greek νψμφη which the anatomist calls nympha or clitoris'.[8] The usage of the term nymphomania thus seems to reflect previous anatomical discoveries about the existence of the clitoris (nympha) and the function of the labia minora (nymphae) since the middle of the sixteenth century.[9] Realdo Colombo, the self-declared discoverer of the clitoris, described it as 'women's seat of lust and enjoyment'.[10] And Tobias Peucer notes:

> The use of the *clitoris* is to wake up venus from her sleep; for its upper part is so very sensitive that it is appropriately called love-apple.... The *nymphae* bring great joy and pleasure in intercourse, they grow bigger and therefore cause a pleasurable titillation when touched by the male member.[11]

But what exactly is nymphomania's relationship to nympha and nymphae? To answer this question I want to take a closer look at the medical descriptions of nympha and nymphae. In 1756 the *Medizinisches Lexicon* offers the following account:

> *Nymphae*, or *water lips*, the inner, small lips of the female pudendum.... They very much resemble the beard or red pieces of flesh which hang from a cock's chin .... It is assumed that they both lead the passage of the urine in such a way that it does not wet the feet and enhance sexual pleasure because of their sensitivity; and both assumptions are very probable.[12]

The *Medizinisches Lexicon* did not invent the strange comparison between nymphae and a cock's wattles. Almost a century earlier, Elias Wallner, in his translation of Bartholin's *Anatomia Reformata*, had noted that they 'resemble the shape of a cock's comb'.[13] Other anatomists, however, saw the nymphae in a very different light. Helkiah Crooke associates them with the world of the water nymphs:

> The nymphae leade the urine through a long passage as it were between two walles, receyving it from the bottome of the cleft as out of a Tunnell: from whence it is that it runneth forth in a broad stream with a hissing noise, not wetting the wings of the lap in the passage; and from these uses they have their name of Nymphes, because they joyne unto the passage of urine, and the neck of the womb; out of which as out of fountaines (and the nymphes are said to bee presedents or dieties of the fountaines) water and humours do issue: and beside,

> because in them are the veneriall delicacies, for the Poets say that the Nymphes lasciviously seeke out the Satyres among the woods and forests.[14]

Peucer refers to the same metaphorical context when he notes that the 'nymphae have received their name because they are situated around the urethral orifice like two water goddesses'.[15] At first glance both metaphorical descriptions seem to be worlds apart, but in fact they are connected by processes of displacement within and between art and medicine during the seventeenth century. At the core of these displacements is the figure of the water nymph.

Nymphs/nymphae

In Greek and Roman mythology water nymphs were virginal goddesses that represented the divine power of water. The *Oxford English Dictionary* defines them as 'semi-divine beings, imagined as beautiful maidens inhabiting the sea, rivers, fountains'. For the function of the nymphs, their ambivalent relation to water is especially important. On the one hand, the water of their dwellings, which was also considered as their gift to humankind, symbolises purity, fertility and sexuality, therefore 'they embody nature's regenerative power and were considered sources of fecundity'.[16] On the other hand, because of its elasticity, water was regarded as a symbol of lascivity and feminine eroticism, and in particular water in motion was known for its masturbatory potential.[17] In Renaissance art, water nymphs figure prominently in pictures, texts, music and architecture (especially fountains), where they convey a sense of idealised and eroticised nature.

Representations of Diana, goddess of chastity and fecundity, bathing with her nymphs were especially popular.[18] Lucas Cranach alone painted at least six different versions of the *Nymph of the Spring*, the most famous of which is now housed in The National Gallery of Art, Washington, DC (Figure 11.1).[19] The sexual symbolism of this painting is ambivalent. The picture shows a beautiful, eroticised female nude sleeping by a fountain; and, although her eyes and pubic hairs are covered by a transparent veil symbolising her chastity, the two partridges at her feet as well as the fountain by which she rests represent sexual lust. And yet, the quiver hanging on the tree seems to suggest that she belongs to Diana's chaste circle. Finally, the inscription at the upper-left part of the picture, which reads, 'Here I sleep, nymph of the spring; do not disturb my sleep' [*Fontis nympha sacri somnum ne rumpe quiesco*] is ambiguous because *rumpere* not only means 'to disturb' (the nymph's sleep) but also 'to tear apart' (her veil) and in this sense the inscription contains a barely disguised invitation to rape. Cranach's painting

11.1 Lucas Cranach, The Elder, *The Nymph of the Spring*, 48.5 x 72.9 cm, oil on canvas, after 1537.

allows for another reading as well: in the context of anatomy's appropriation of the nymphs to describe both the female external genitals and women's erotic pleasure, the representation of the spring might be regarded as a visualisation of the anatomical nature of the nymphae *before* anatomy. In fact, Crooke's description of the nymphae reads like a version of Cranach's painting, if one substitutes spring for nymphae and water for urine: 'the nymphae [the spring] leade the urine [water] through a long passage as it were between two walles, receyving it from the bottome of the cleft as out of a Tunnell: from whence it is that it runneth forth in a broad stream with a hissing noise'. Envisaged in this way, the connection between anatomy's nymphae and the mythological world of the water nymphs fabricated by Crooke and Peucer is more than a metaphorical reference/reverence. Connecting water nymphs and the female genitals was not confined to the realm of anatomy. In Martin Opitz's famous 'Schäfferey von der Nimfen Hercinie' the narrator remarks that 'whosoever drinks from the clitoral fountain shall not like the smell even of wine anymore'.[20]

The mythological figure of the water nymph, as role model for the mapping of female sexuality, is more than a topos. Rather, these borrowings point to a collective cultural fantasy about the nature of female sexuality. In other words, anatomy's appropriations suggest that the nymphs figured as the embodiment of the seemingly forgotten knowledge about the anatomy of female erotic pleasure during the Middle Ages. However, the introduction of the nymphs into anatomical science changed their nature significantly: during the seventeenth century the erotic pleasures of the nympha(e) came to be different from those of the nymphs. Despite the striking similarities between Crooke's anatomical rendering of the nymphae and Cranach's *Nymph of the Spring*, the differences are clear. Crooke emphasises the heterosexual pleasure of the nymphae, comparing them to the 'nymphs lasciviously seek[ing] out the satyres among the woods and forests'.[21] In Cranach's painting the nymph's heterosexuality is neither obvious nor necessary. It is not clear at all *whose* genitals are being represented in the painting and several readings are possible. First, the spring represents the nymph's own pudendum. In this view the picture displays her sexual autonomy and self-sufficiency as she enjoys the stimulation of the erotically charged water running past her nymphae. Second, the spring might be read as the sleeping nymph's desire for another woman's nymphae. In this reading, the picture stages a homoerotic fantasy. Third, if the viewer of this picture is a woman her gaze may touch the body and genitals of the sleeping nymph. Thus in Cranach's picture the representation of the nymph's sexuality exceeds a male heterosexual claim on female sexuality. This is, of course, not to say, that Cranach intentionally painted a masturbatory, even 'lesbian' picture. Rather, as Patricia Simons has noted, 'the very sensual anarchism or slipperiness of the visual image may encourage *deviant* and *perverse* possibilities we have not yet allowed ourselves to see'.[22] Connecting Cranach's nymph to the anatomists' nymphae may seem 'perverse', but considering Crooke's and Peucer's appropriation of the nymphs for their own anatomical writing the 'perversity' of my reading reiterates the 'perversity' of early-modern science. And Cranach's picture is no exception to the rule – numerous Renaissance paintings focusing on mythological scenes from the life of Diana and her nymphs are even more explicit.

According to ancient Greek and Roman mythology, Diana lived with her nymphs in a chaste community where men were not permitted. As Christine Downing has noted, Diana 'is the goddess most intimately associated with female embodiment' and sexuality.[23] The story of Actaeon who accidentally came upon the grotto where Diana and her nymphs were bathing and was immediately transformed into a stag, and the story of the nymph Callisto who was expelled from the circle when her pregnancy resulting from a rape by Jupiter was discovered, both emphasise the autonomous and homoerotic sexuality in Diana's band.[24] These stories were extremely popular during the early-modern period and there exist a great number of pictorial, verbal and musical representations, some of which 'could so stress their [the nymphs'] exclusive femininity that marriage and maternity were explicitly disavowed'.[25] A sixteenth-century painting clearly emphasises this sexually charged, auto- as well as homoerotic atmosphere. Although art historians have claimed that the image shows Diana expelling Callisto from her train, this title was only added in the nineteenth century.[26] In contrast to Titian's well-known rendering of the subject, in this painting the discovery of Callisto's pregnancy does not seem important to the nymphs; with the exception of Diana, no other woman seems to be interested in her condition. Two nymphs are enjoying their bath in the mighty waterfall, which clearly is the centre of the spacious grove, while others are either getting ready for or have just finished their bath. Diana herself, who is being tenderly embraced by one of her companions, seems more concerned than angry. Moreover, the nymphs enjoy touching themselves: to the right, one of the nymphs caresses her pudendum with a cloth, while another one, whose face we are not allowed to see, masturbates at Diana's feet.

Thomas Heywood's play *The Golden Age*, first performed in 1609, contains another rendering of the Diana and Callisto story that features homoerotic desire between women.[27] When Callisto is admitted to the circle, Diana calls for a 'princess in our train, / as yet unmatch'd to be her cabin fellow, / and sleep by her' (Act II, scene 1).[28] When she is told by her handmaid that they 'all are coupled / and twinn'd in love, and hardly is there any / that will be won to change her bedfellow', Diana rules that she must wait 'till the next arrive: She that is next admitted of our train, / must be her bed-companion'. When Jupiter, who has arranged to be next, joins the circle in the guise of a woman, Diana's handmaid advises her to abstain from heterosexual intercourse by all means:

> You never shall with hated man atone
> But lie with woman, or else lodge alone …
> With ladies only you shall sport and play,
> And in their fellowship spend night and day …
> Consort with them at board and bed,
> And swear no man shall have your maidenhead.

As Valerie Traub has noted, Heywood's representation of the story 'poses "sport and play" between women as a chaste alternative to penetrative sex with "hated man"'.[29] Jupiter's rape of Callisto thus appears as an act of violence and aggression which sharply contrasts with 'the loving bonds between women'.[30] As far as female homosexuality in Diana's circle is concerned, Joachim Wtewael's 1607 painting *Diana and Actaeon* could not be more explicit (Figure 11.2). In this image Actaeon disturbs two nymphs who are engaging in sexual practices: one woman tenderly strokes the inside of the other's thighs while at the same time her right hand caresses her own breast.

The drama and paintings discussed here do not condemn the nymphs' masturbatory practices and homoerotic pleasures, although masturbation was regarded a sin and female homosexuality was a crime in most European countries. As Simons and Traub suggest, the discourse on chastity to which these pictures belonged served as a 'veil' under which female sexuality that did not fulfil heterosexual prerogatives could be represented and consumed – for a while, at least. Although anatomists appropriated without hesitation the sexually-charged world of the nymphs for their anatomical mapping of female sexuality, they also significantly modified that sexuality. Correggio's painting *Leda and the Swan* (1530–32) is a striking example of how anatomy silently altered the world of the nymphs (Figure 11.3). Referring to Euripides' *Helena*, the picture represents a particularly sexist version of Leda enjoying being raped by Jupiter in the guise of a swan. I do not want to expand here on the mythological aspects of this painting; nonetheless, it provides important insights into the history of the clitoris.

'As long as a swan's neck'

Although anatomists celebrated the discovery of female erotic pleasure, they also grew anxious about its sexual possibilities.[31] Anatomy books report cases of monstrously enlarged nymphae and clitorises that allowed women to imitate men, and seduce and have sex with other women. Peucer notes that both nymphae and clitoris might grow to such enormous lengths that they

11.2 Joachim Wtewael, *Diana and Acteon*, 58 x 79 cm, oil on wood, 1607.

11.3 Antonio Allegri Correggio, *Leda and the Swan*, 152 x 191 cm, oil on canvas, *c.* 1530–32.

obstruct intercourse and thus have to be excised. He adds, however, that an enlarged clitoris is much more 'inconvenient for congress' than an enlarged nymphae because it is harder than the nymphae.[32] In fact, the clitoris was regarded as a female penis that, as Jane Sharp notes, 'will stand and fall as the yard doth and makes women lustful and take delight in copulation'.[33] For some anatomists like Crooke or Bartholin an enlarged clitoris was particularly troublesome because they regarded it as the locus of female homosexuality. As Bartholin states: 'sometimes they abuse the clitoris as if it were a penis and they lie with other women'.[34] It is because of this homoerotic disposition that Bartholin and many of his colleagues refer to the clitoris as 'contempt of men'.[35] By the beginning of the seventeenth century, this 'contempt of men' was made culturally intelligible in the monstrous figure of the 'tribade' – a woman who because of her enlarged clitoris desired and had sex with other women. Crooke, for instance, describes her as an abuser of the clitoris: 'And this part it is which those wicked women do abuse called Tribades (often mentioned by many Authors, and in some states worthily punished) to their mutuall and unnaturall lustes.'[36] As Traub notes, 'It is not the "tribade's" inconstant mind or sinful soul but her uniquely female yet masculinized morphology that propels her to engage in illicit behavior'.[37] As a result doctors recommended nymphotomia (that is, the excision of nymphae and/or clitoris) as a measure to ensure normal, heterosexual intercourse between men and women.[38]

How did anatomists see, or rather imagine, these enlarged female genitals? As Bartholin writes: 'It is absolutely true and it is not natural and it is monstrous that it grows to the length of a goose's neck.'[39] Bartholin's remark refers to a case described by the Swiss anatomist Felix Platter whose observation soon became a popular topos in the anatomical writing about the clitoris. Peucer even reports 'Platter testifies to having seen one as long as a swan's neck'.[40] In this context it is striking that, although the visual representation of the body's structure and interior was an important feature of early-modern anatomy, books do not contain illustrations of enlarged nymphae or clitorises.[41] However, Correggio's depiction of Leda, albeit painted before the anatomical rediscovery of the clitoris, gives us a visual impression of the clitoris's sexual and erotic possibilities, and carries a suggestion that the nymphs embodied knowledge of the female genitals and female erotic pleasure. But, while the anatomical discourse fashioned the enlarged clitoris and the desire it instilled into a monstrous thing, and while doctors recommended nymphotomia in order to discipline female sexuality, in Correggio's *Leda and the Swan* no such condemnation (or operation) is implied. On the contrary, as in the case of Diana and her nymphs, the mythological content seems to cover up for the homoerotic dimension. In other words, the appropriation of the nymphs by the anatomical discourse must be considered as a normalising gesture that fabricates heterosexual practices as the norm and homosexual practices as deviant. Obviously anatomy found inspiration in art. The question, then, is where does this interest in the clitoris and nymphae in both art and science come from and what is the significance of the negotiations between art and science?

It is important to keep in mind that the deconstruction of the nymphs by anatomy cannot be regarded as a conscious act of the individual anatomist. Rather, the treatment of the nymphs is part of the 'culture of dissection' that Jonathan Sawday has described. For Sawday, dissection means more than the fragmentation of the corpse by the anatomist, indeed, he takes it to describe a social and cultural activity that includes a 'violent reduction of parts: a brutal dismemberment of people, things, or ideas'.[42] The term 'culture of dissection' suggests that anatomy, with its new focus on dismemberment and partition, is at the centre of this 'network of practices, social structures and rituals' and Sawday identifies *Anatomia*, the goddess of division, as 'the emblem of this culture'.[43] It is, however, important to note that anatomical dissections did not *cause* a different perception of the female genitals. After all, anatomists had performed dissections of the female body since the end of the thirteenth century, albeit not quite as frequently as in the early-modern period, without 'discovering' the nature of female sexual pleasure.[44] When more than 200 years later the meaning of the female genitals, as well as the nature of female erotic pleasure, became so important to anatomy, it was not because anatomists for the first time since antiquity had opened up and looked into the female body, but because they *saw* this body *differently*. As we shall see, the reintroduction, and improvement, of linear perspective into European culture at the end of the fifteenth century led to new perceptions of female sexuality. The culture of dissection is not an effect of anatomy, rather anatomy became so important because it offered the opportunity to provide a body for this new vision.

The heterosexual vision

As Marshall McLuhan has argued, 'the message of any medium or technology is the change of scale or pace or pattern that it introduces into human affairs'.[45] What was linear perspective's contribution? First, perspective is a medium providing at once the 'correct' *perception* as well as the 'correct' *representation* of the world.[46] This vision is achieved by a reduction of the world to the point of view of the one-eyed, immobile, sovereign and detached, male viewer and its counterpart, the vanishing point. Albrecht Dürer in his *Entwürfe* explained that 'the division of things that one sees' was crucial to achieve the 'right vision'.[47] This division could be manufactured with the help of the so-called 'velum', a grid made of very fine, transparent cloth put into a frame through which the artist looked at his object. Walter Ryff described the velum as a 'pure, thin … fabric of the most unadulterated fibre'.[48] The perspectival gaze produced dichotomies – for example: active/passive, male/female, subject/object – and thus was the medium in and with which the culture of dissection came to be realised. Linear perspective put the anatomist in a position from which to determine the 'truth' about the human body – after all anatomical dissections of humans were first performed for legal purposes[49] – and it is anatomy that provided flesh for the geometrical patterns of perspective. Renowned artists like Leonardo da Vinci and Titian, Andrea del Sarto or Domenico del Babiere, who experimented with perspective, attended dissections and were responsible for the illustrations in many of the famous anatomy books.[50]

What is true for the body is also true for sexuality. Linear perspective not only divides the visible world into binary oppositions, it also envisions seeing in terms of heterosexual reproduction. Teachers of perspective over and over again emphasised that the truthful representation of the visible world was based on images conceived in the mind, the so-called '*concetto*'.[51] As Dürer noted, a good painter is full of concepts which he 'ejaculates' in and through his work and this will grant him immortality.[52] Linear perspective allowed the artist – like God – to create the visible world by a pure act of will, and perspectival rules were meant to help the artist impregnate the unformed material with his own concepts. Consequently, the world perceived, as well as the images produced, were proof of the artist's potency. Thus, looking becomes a form of heterosexual intercourse and the eye a transmitter of sperm. Consider Aristotle's influential assumptions about the different roles and functions of men and women for the *Generation of Animals*: 'male and female differ in their logos, because the male is that which has the power to generate in another … while the female is that which can generate in itself, i.e. it that out of which the generated offspring, which is present in the generator, comes into being'.[53] And, he continues: 'The male provides form and the principle of the movement, the female provides the body, in other words the material.'[54] In his *Disputations Touching the Generation* (1657) William Harvey stated that artistic creation and heterosexual reproduction were in fact one and the same for 'the generation of things in Nature and the generation of things in Art take place in the same way… Both are first moved by some conceived form which is immaterial and is produced by conception.'[55]

Envisaged in this way, the velum, this 'purest' and 'thinnest' of all fabric, functions as an artificial hymen awaiting the penetration of the artist's gaze. Within this heterosexual matrix, sexuality is visualised and vision is sexualised. With the development and circulation of perspective in Europe, gender and sexuality were increasingly represented in terms of (male) looking and (female) being looked at.[56] In this context the renaissance of nymphs in paintings, which took place from the second half of the fifteenth century onward, reflects perspective's focus on sexuality and reproduction. The nymphs represent the masculine concept of self-generation and sexualised vision while at the same time, some of these pictures resist heterosexuality and thus invite a queer gaze.[57] However, with the increasing visualisation of sexuality during the sixteenth and seventeenth centuries, this particular kind of female sexual self-sufficiency and autonomy slowly disappears from both the canvases and anatomy. These displacements include the boisterous condemnation as well as the silent effacement of female homosexuality.

As Traub has shown, over the course of the seventeenth century, dramatists significantly change the meaning of the nymphs' sexuality in Diana's band. While, at the beginning of the century, Thomas Heywood depicted the nymphs' homoeroticism as natural, in later dramas this desire is staged precisely in order to be condemned. However, Callisto is not banished because she had sex with a man but because she had sex with a woman; Diana and her nymphs are chaste precisely *because* they are heterosexual yet virginal. This transformation is accompanied by a modification of the nymphs' environment, that is, of their sexual and erotic nature. In this context water is of special importance and in paintings these alterations are most clearly visible.

While in earlier paintings water is abundant, later paintings show the gradual drying up of their grove.[58] This dryness is significant because, as I have indicated, water, especially running water, symbolised the sexual energies of the nymphs and was central to their nature; hence, the lack of water in later paintings implies an alteration of this nature. In Louis de Boullogne's *Diana Resting* (1707) the water in the nymphs' grove has been reduced to a little puddle useful for nothing more than a footbath (Figure 11.4). Sometimes there is so little water it is hard to speak of a spring at all. The increasing dryness in these paintings also suspends the association between nymph and nymphae or clitoris. At the same time, clothes become more important: in contrast to the earlier paintings I have discussed, later pictures show Diana and her nymphs either fully dressed or draped with cloth in such a way that direct skin contact between them is prevented. While these paintings certainly represent erotic relationships between women, they do so by focusing on their chaste rather than on their sexual dimension. To put it differently, while in earlier paintings the discourse of chastity served as a veil to represent and consume female homoeroticism, in later paintings female homoeroticism is represented *as* chaste.[59]

Another version of the nymphs' chastity, this time by keeping the water flowing, can be found in Opitz's aforementioned *Schäfferey*. In this text, three

11.4 Louis de Boullogne, *Diana Resting*, oil on canvas, 1707.

poets, while on a journey across the *Riesengebirge* during the Thirty Years War, all of a sudden encounter the nymph Hercinie 'who in a fresh grotto or cave lay on her left arm … dressed in a subtle and transparent veil, her hair adorned with a green wreath and draped in a foreign fashion, and in her right hand she held a container of very white marble out of which ran a small creek'.[60] Unlike Actaeon, who is severely punished for surprising the nymphs at their bath, the travellers are greeted with a song and, what is even more surprising, they are invited to visit the grotto. Upon entering the grotto, the three poets are impressed by the waterfalls: 'When we came into the cave, we could not see anything else but water of the purest kind … we found ourselves in a very cool grotto out of which not only this water came running, there were other streams surfacing through hidden paths and arteries of the rocks.'[61] Surprised and overwhelmed, they 'almost lost their hearing because of the loud noise and running of the cataract'. In another room – which 'usually remains closed for male eyes', as Hercinie tells them – they encounter nymphs occupied with 'spinning, weaving and sewing'.[62] Finally, they set foot in the inner sanctuary of the grotto: a room dedicated to the memory of the Schaffgotsch family's male heirs who have secured and protected the wealth and freedom of the *Riesengebirge* and who are therefore honoured and cared for by the nymphs. As Hercinie explains to the visitors: 'Their ancient blood, their virtue, their praiseworthy deeds and especially the quiet life which we have enjoyed with them as our guardians has earned them the erection of this memorial in our midst.'[63] In Opitz's text the nymphs gather around male heroes whose accomplishments they commemorate and honour; their lodgings are characterised by male, not female potency. Thus, in the *Schäfferey* men become acquainted with the reassuring truth that the nymphs' grotto actually is a sanctuary of male potency and fertility.

After having resurfaced from the grotto the poets continue their journey in search of the so-called 'warm fountain' whose water is said to possess magic healing powers.[64] Especially interesting is the description of the quality of its water: 'In their midst there was the famous fountain with its spouting water generating a lot of tiny bubbles of a clear, transparent colour like a white saffire with a little bit of blue.'[65] This kind of water strikingly resembles descriptions of male semen. According to Aristotle's influential version,

> Semen, then, is a compound of pneuma and water (pneuma being hot air), and that is why it is fluid in its nature, it is made of water. Semen is thick and white.… The cause of the whiteness of semen is that it is foam, and foam is white, the whitest being that which consists of the tiniest particles, so small that each individual bubble cannot be detected by the eye.[66]

Harvey's observation that male sperm is 'permeated with spirit by the fervency of coitus or desire and froth with the nature of spume' clearly echoes

Aristotle's definition.[67] In the *Schäfferey* the nymphs of the spring do not possess any of the sexual autonomy their predecessors had. Instead of female fecundity and sexuality their water now symbolises male fertility and instead of being honoured themselves they exist to honour the male benefactor who erected the fountain or protects the grove.

This same tendency to heterosexualise female sexuality can be observed in the anatomical discourse where, by the end of the seventeenth century, the nymphae and clitoris do not resemble Cranach's self-contained nymph or Correggio's aggressive swan but, as I mentioned at the beginning of this essay, 'a comb of a cock'.[68] Rather than possessing sexual pleasures of their own, clitoris and nymphae have been transformed into a kind of adornment serving to arouse women's and men's lust for heterosexual intercourse. It is at this point in history that the term 'nymphomania' is introduced in the anatomical and medical discourse. As the pathological 'other' of the chaste and heterosexual nymph, the nymphomaniac woman embodies the male fantasy of a woman whose well-being depends upon sexual satisfaction provided by men. At the end of the seventeenth century, both nymph and nymphomaniac refer to an invisible male voyeur whose potency and fertility they embody and make visible. The introduction of the term nymphomania marks precisely this transformation of the nymphae and clitoris into heterosexual and yet inconsequential genitalia.

The extent to which the heterosexual vision tends to naturalise the bodies and sexualities it helped to construct becomes evident in Sigmund Freud's theories on the development of femininity. In his 1931 essay 'Female sexuality' he writes:

> We have long understood that the development of female sexuality is complicated by the fact that the girl has the task of giving up what was originally her leading genital zone – the clitoris – in favour of a new zone – the vagina. ... Thus in female development there is a process of transition from one phase to the other, to which there is nothing analogous in the male.[69]

This shift of the girl's erotogenic sensibility from the clitoris to the vagina is caused by her 'acknowledgement' that she lacks the penis and thus 'grows dissatisfied with her clitoris'.[70] Athough Freud was certain he had found the anatomical foundation for the development of female heterosexuality, he had to admit that he was not yet able to describe the biology of this development: 'We do not, of course, know the biological basis of these peculiarities in women.'[71]

In the light of the history of the nymphs and nymphomania, it seems obvious that Freud was not able to find an answer to this question because he was looking for *biology* where there is *culture*. As I have tried to show, the introduction of nymphomania into medical and anatomical discourse indicates that the shift from clitoris to vagina, from auto- or homoeroticism to full womanhood, as Freud would have it, is in fact a *cultural* and *historical* process. This process tells us about the importance of visuality for female erotic pleasure – a relation that Freud fully acknowledged. In both Freud's and my own account, sexuality and vision cannot be separated. In 'Some psychical consequences of the anatomical distinction between the Sexes', Freud attributes women's erotic abandonment of the clitoris to the frustration the little girl experiences when she 'notice[s] the penis of a brother or a playmate, *strikingly visible* and of large proportions, [and] at once recognizes it as the superior counterpart of [her] own small and *inconspicuous* organ, and from that time forward fall[s] a victim to the envy for the penis'.[72] The little boy, on the other hand, 'begins by showing irresolution and lack of interest' when he 'first catches sight of a girl's genital region'.[73] However, in fact it is the male's eagerness to gaze on the female genitals that gave rise to the cultural fantasy 'that the elimination of clitoridal sexuality is a necessary precondition for the development of femininity'.[74] From this perspective, women's envy for the penis appears as a projection (in every sense of the word) of male clitoris-envy: 'It is absolutely true and it is not natural and it is monstrous that it grows to the length of a goose's neck.'[75]

Notes

Unless otherwise indicated, all translations from the German or Latin are my own. I am grateful to Uta Scheer and Samuel Willcocks for redirecting my gaze, and to Caroline Scherf for Freudian services. The writing of this paper was supported by a research grant from the Berliner Senatsverwaltung für Arbeit, Soziales und Frauen.

1. Hayd, J. (1968), *Die Hysterie und Nymphomanie dargestellt in Doktorarbeiten des 17. und 18. Jahrhunderts aus der Dr. Heinrich Laehr Sammlung*, München (dissertation), p. 54.
2. Haller, A. von (1756), *Onomatologia Medica Completa oder Medizinisches Lexicon*, Ulm, Frankfurt and Leipzig: Saumische Buchhandlung; Woydt, J.J. ([1701] 1729), *Gazophylacium Medico-Physicum oder: Schatzkammer Medicinisch-und natuerlicher Dinge*, 6th edn, Leipzig: Friedrich Lanckisches Erben.
3. Cited in Hayd (1968), p. 55.
4. Haller (1756), p. 684.
5. Woydt ([1701] 1729), p. 379.
6. Ibid., p. 380.
7. Maines, R.B. (1999), *The Technologies of Orgasm. 'Hysteria', the Vibrator, and Women's Sexual Satisfaction*, Baltimore and London: Johns Hopkins University Press, pp. 1–7.
8. Cited in Hayd (1968), p. 54.
9. Traub, V. (1995), 'The Psychomorphology of the Clitoris'. *GLQ*, **2**, 81–113.

10. Colombo, R. (1609), *Anatomia*, trans. I.A. Schenkius, Frankfurt-am-Main: Eggenolff, p. 204.
11. Blancaert, S. (1691), *Reformirte Anatomie*, trans. T. Peucer, Leipzig: Weidmann, pp. 817–18, my emphasis.
12. Haller (1756), p. 1154.
13. Bartholin, T. (1677), *Neu-verbesserte Kuenstliche Zerlegung deß Menschlichen Leibes*, trans. E. Wallner, Nürnberg: Hoffmann, p. 311.
14. Crooke, H. (1651), *Microcosmographia: A Description of the Body of Man*, London: R.C., pp. 237–8.
15. Blancaert (1691), p. 817.
16. *Paulys Realencyclopaedie der Classischen Altertumswissenschaften* (1937), ed. W. Kroll, Stuttgart: Metzler, **34–35**, p. 1536.
17. Bredekamp, H. (1988), 'Wasserangst und Wasserfreude in Renaissance und Manierismus', in Böhme, H. (ed.), *Kulturgeschichte des Wassers*, Frankfurt-am-Main: Suhrkamp, pp. 145–88; Maines (1999), p. 12.
18. Reid, J. (1993), *Classical Mythology in the Arts, 1300–1990s*, 2 vols, New York and Oxford: Oxford University Press.
19. Koepplin, D. and Falk, T. (1974), *Lukas Cranach. Gemälde, Zeichnungen, Druckgraphik*, 2 vols, Basel and Stuttgart: Birkhäuser Verlag, pp. 631–40.
20. Opitz, M. (1990), 'Schäfferey von der Nimfen Hercinie', *Gesammelte Werke. Kritische Ausgabe*, 6 vols, ed. G. Schulz-Behrend, Stuttgart: Anton Hiersemann, vol. 4, Pt. 2, p. 572.
21. Crooke (1651), p. 237.
22. Simons, P. (1994), 'Lesbian (In)Visibility in Italian Renaissance Culture: Diana and other cases of *donna con donna*', *Journal of Homosexuality*, **27**, 81–122 (p. 110).
23. Downing, C. (1989), *Myths and Mysteries of Same-Sex Love*. New York: Continuum, p. 210.
24. Simons (1994), p. 107.
25. Ibid., p. 97.
26. Cartari, V. (1996), *Imagini de i Dei de gli Antichi*, Vicenza: Pozza, p. 156.
27. Traub, V. (1996), 'The Perversion of "Lesbian" Desire', *History Workshop Journal* **41**, 23–49.
28. Heywood, T. (1966), *Dramatic Works: The Golden Age*, Nendeln: Kraus.
29. Traub (1996), p. 29.
30. Ibid., p. 29.
31. Park, K. (1997), 'The Rediscovery of the Clitoris. French Medicine and the Tribade, 1570–1620', in D. Hillman and C. Mazzio (eds), *The Body in Parts. Fantasies of Corporeality in Early Modern Europe*, New York and London: Routledge, pp. 171–93.
32. Blancaert (1691), p. 310.
33. Sharp, J. (1999), *The Midwives Book, Or the Whole Art of Midwifery*, ed. E. Hobby, New York and London: Oxford University Press, p. 40. On Sharp's text generally, see also Hobby's chapter in this volume (Chapter 9).
34. Bartholin, T. (1660), *Anatomia ... Reformata*, Hagae-Comitis: Adriani Vlacq, p. 186.
35. Ibid., p. 186.
36. Crooke (1651), p. 176.
37. Traub (1995), p. 94.
38. Park (1997), p. 184.

39. Bartholin (1660), p. 186.
40. Blancaert (1691), p. 816.
41. I have argued elsewhere that the visual representation of an enlarged clitoris would have collided with the phallus of the gaze; see my paper 'As Long as a Swan's Neck: The Significance of the Lesbian Clitoris for Early-Modern Anatomy', given at the conference titled 'Virile Women, Consuming Men: Gender and Monstrous Appetites in the Middle Ages and Renaissance', University of Wales Conference Centre, Grgynog, 25–27 April, 2000.
42. Sawday, J. (1995), *The Body Emblazoned. Dissection and the Human Body in Renaissance Culture*, London and New York: Routledge, p. 1.
43. Ibid., pp. 2–3.
44. Siraisi, N.G. (1990), *Medieval and Early Renaissance Medicine. An Introduction to Knowledge and Practice*, Chicago: University of Chicago Press, pp. 86–97.
45. McLuhan, M. (1987), *Understanding Media. The Extensions of Man*, London and New York: Ark, p. 8.
46. Busch, B. (1997), *Belichtete Welt. Eine Wahrnehmungsgeschichte der Fotografie*. Frankfurt-am-Main: Fischer, pp. 61–88.
47. Dürer, A. (1995), 'Entwürfe', in T. Kramer and C. Klemm (eds), *Renaissance und Barock*, Frankfurt-am-Main: Deutscher Klassiker Verlag, p. 86.
48. Ryff, W. (1995), 'Unterrichtung zu rechtem Verstandt der lehr Vitruuij', in Kramer and Klemm, pp. 117–301, p. 145.
49. Siraisi (1990), p. 86.
50. Roberts, K.B and Tomlinson, J.D. (1992), *The Fabric of the Body. European Traditions of Anatomical Illustrations*, Oxford: Clarendon Press.
51. Link-Heer, U. (1986), 'Maniera: Überlegungen zur Konkurrenz von Manier und Stil (Vasari, Diderot, Goethe)', in H.U. Gumbrecht and K.L. Pfeiffer (eds), *Stil: Geschichten und Funktionen eines kulturwissenschaftlichen Diskurselements*, Frankfurt-am-Main: Suhrkamp, pp. 93–114.
52. Dürer (1995), p. 83.
53. Aristotle (1958), *Generation of Animals*, trans. A.L. Peck, Cambridge, MA and London: Harvard University Press, p. 13.
54. Ibid., p. 109.
55. Quoted in Laqueur, T. (1990), *Making Sex. Body and Gender from the Greeks to Freud*, Cambridge, MA and London: Harvard University Press, p. 147.
56. Hammer-Tugendhat, D. (1994), 'Erotik und Geschlechterdifferenz. Aspekte zur Aktmalerei Tiznias', in D. Erlacher, M. Reisenleitner and K. Vocelka (eds), *Privatisierung der Triebe. Sexualität in der Frühen Neuzeit*, Frankfurt-am-Main: Peter Lang, pp. 367–446.
57. Joseph Haintz's picture *Diana and Actaeon* (*c.* 1590) stages such a queer gaze.
58. See, for example, Antoine Coypel, *Diana bathing* (1695) and François Boucher, *Jupiter seducing Callisto* (1759).
59. Simons (1994) and Traub (1995).
60. Opitz (1990), p. 533.
61. Ibid., p. 534.
62. Ibid., p. 535.
63. Ibid., p. 540.
64. Ibid., p. 571.
65. Ibid., p. 575.
66. Aristotle (1958), p. 163.
67. Cited by Laqueur (1992), p. 146.

68. Bartholin (1677), p. 311.
69. Freud, S. (1990a), 'Female sexuality', in *The Major Works of Sigmund Freud*, Chicago: Enyclopedia Britannica, Great Books of the Western World series, No. 54, pp. 371–4.
70. Ibid., p. 376.
71. Ibid., p. 374.
72. Freud, S. (1990b), 'Some psychical consequences of the anatomical distinction between the Sexes', in *The Major Works of Sigmund Freud*, Chicago: Encyclopedia Britannica, p. 335.
73. Ibid., p. 336.
74. Ibid., p. 339.
75. Bartholin (1660), p. 186.

PART IV
NEW WORLDS AND NEW PHILOSOPHIES

CHAPTER TWELVE

Thomas Harriot and John White: Ethnography and Ideology in the New World

Andrew Hadfield

How scientific were European attempts to classify peoples in the sixteenth and seventeenth centuries? The problem has a particular significance because, while such classification was not a new phenomenon, the recent discovery of the Americas posed fundamental questions about the nature of the peoples encountered there, and, consequently, mankind in general. As has often been pointed out, the main sources for medieval anthropology were Herodotus' *Histories* (fifth century BC) and Pliny the Younger's *Natural History* (first century AD). It was generally believed that all the peoples of the world created by God were known to Christian Europe either through direct encounter or eye-witness accounts. A thorough search through the works preserved in libraries would reveal accounts of all of God's people.[1] Works such as the Mappa Mundi, preserved in Hereford Cathedral, divide the world neatly into a tripartite structure, half of the circle being taken up by Asia, and the other half split neatly into Europe and Africa. All known peoples are contained within the circle of the world; strange and bizarre creatures of uncertain human status, such as the men whose heads grow beneath their shoulders, and the sciapods, who find shelter under their huge feet, inhabit the shadowy margins.[2]

Such faith was severely altered by the discovery of the New World in 1492. Columbus clearly interpreted the peoples he encountered in terms of those he had already read about, believing that he had reached the shores of the lands of the Great Khan described in Marco Polo's *Travels*, and thinking that the man-eating creatures he was told about by the natives he dealt with were the anthropophagi referred to in Herodotus.[3] And, for many writers, the models and means of classifying peoples they had inherited remained securely in place. Jean Bodin, for example, makes no mention of the Americas in the geographical survey contained in his *Method for the Easy Comprehension of History* (1566).[4] But, for others, the discovery of the natives of the Americas challenged the whole basis of their knowledge. Stephen Greenblatt has characterised the fundamental European response as one of 'wonder' at the strangeness and unfamiliarity of the encounter, 'the decisive emotional and intellectual experience in the presence of radical difference'.[5]

In Spain, the principal European nation involved in travel to and colonisation of the Americas until the early seventeenth century, intellectual efforts to analyse and comprehend who and what had been discovered led to fundamental disagreement and argument. The natives were regarded by many as the 'natural slaves' of which Aristotle had written, barbarians who had failed to progress, as they should have done, towards Christianity and civility. They were naturally inferior to civilised, Christian Europeans and so could be used as their superiors saw fit because they would never attain the status of full humanity. The freedom they enjoyed as naked people outside the constraints of the law evaporated as soon as they came into contact with civilisation.[6] Others argued that the natives were, in Aquinas's terms, 'invincibly', rather than 'vincibly' ignorant like Jews and Muslims, who had heard the word of God, but had resisted conversion.[7] The Americans were 'nature's children', part of God's family who had simply not had a chance to hear the word of God yet. It was a great opportunity for the Spanish to convert them and so enlarge Christ's empire – and their own. The problem that needed explanation was why God had chosen to neglect a large portion of his children.[8] The dispute culminated in the famous debates at Valladolid in 1550–51, between Juan Ginés de Sepúlveda, who produced the most uncompromising argument for the inferiority of the native Americans, and Bartolomé de Las Casas, the champion of their rights and exposer of Spanish cruelty in the Americas.[9]

This crucially important debate indicates that Greenblatt is oversimplifying in characterising European response to the Americas as defined by 'wonder', but does vindicate his suggestion that the discovery of the New World disrupted European intellectual and emotional certainty when confronted by the experience of 'radical difference'. It also demonstrates the mediation involved between the use of a system of classification already in place, and the shock of the new.[10] Early-modern ethnology, as outlined by Pagden and Hogden, was never free from other modes of discourse or means of classifying people.[11] It was always part of a wider network of analysis, what Colin Kidd has termed 'ethnic theology'. Analysing the variety of British identities before the rise of nationalism in the eighteenth century, Kidd points out that the fundamental narrative of human diversity accepted in the sixteenth century was still the story of the postdiluvian diaspora of Noah's sons:

To appreciate the discursive priorities of the clerics and literati of the early

> modern era who engaged with the issues of ethnicity, it is necessary to liberate the historical imagination. Otherwise the pursuit of ethnicity remains trapped within modern categories ... Within the Mosaic scheme, difference mattered less than degrees of consanguinity among a world of nations descended from Noah. Indeed, the primary value of ethnicity was not ethnological in the modern sense, but lay within the theology of 'evidences', where it functioned as a vital weapon in the defence of Christian orthodoxy and the authenticity of Scripture from heterodox assaults.[12]

No identity could be established without recourse to a theological explanation, a relationship that the Spanish debates about the peoples of the New World neatly illustrates. Kidd's analysis of early-modern British identities in terms of conceptions of 'ethnic theology' also suggests that we should be reading British attempts to classify the peoples they encountered in the Americas in a similar light. There was no elaborately staged debate in Protestant England to match that of Catholic Spain in Valladolid; however, as I shall argue in this essay, ethnological representations of native Americans were conceived along similar lines.

The most frequently reproduced English analysis of native Americans from the late sixteenth century is Thomas Harriot's *A Briefe and True Report of the New Found Land of Virginia.*[13] The *Report* was first published in 1588, in an undistinguished-looking quarto produced by Robert Robinson. It was reprinted with minor editorial changes in 1600 as part of Richard Hakluyt's *The Principall Navigations, Voyages, Traffiques & Discoveries of the English Nation*, and subsequently included in the expanded version of that collection in 1600, indicating that from its first appearance the *Report* was recognised as a key work on the Americas.[14] The most important, widely-circulated and handsomely-produced edition was the folio of 1590, published by the exiled Belgian printer Theodor De Bry as the first part of his multi-volume collection of New World voyages, *America* (1590–1634). Much care and attention was lavished on this edition, the first produced by a major European publishing enterprise that was designed to promote the case for Protestant colonisation of the Americas as a means of confronting and preventing a Spanish, Catholic monopoly of the continent,[15] and it was dedicated to Sir Walter Raleigh, as originator of the Roanoke colony. The *Report* was separately published in four languages, English, Latin, French and German. It is likely that De Bry had been persuaded to publish Harriot's work as the first volume of the series by Richard Hakluyt. The two had met in London in 1587, when De Bry had been in pursuit of Jacques Le Morgues de Moyne, the official artist of the ill-fated French colony in Florida (1562–65), with the intention of using the latter's drawings to accompany the text of René de Laudonniere's account of the enterprise.[16]

Appended to De Bry's edition of the *Report* were a series of 29 drawings, amongst the most sophisticated and best-produced in an English text in the sixteenth century. These were all – with one exception – based on the drawings of John White, who had sailed to the Roanoke colony with Sir Richard Grenville's expedition in 1585, and who had been made governor of the colony in 1587 before returning to England to plead on behalf of the beleaguered colonists for more supplies and support.[17] The captions to the illustrations were written by Harriot in Latin, and translated into English by Richard Hakluyt, further indicating the careful and deliberate planning that went into the production of the edition and alerting readers to the importance of the elaborate detail it contains, verbal and iconographic.[18]

Harriot's actual text remains virtually the same in all its editions, but the significance clearly alters according to the context – a vital consideration when assessing its scientific and anthropological status. Robinson's edition appears to have an exhortationary function, displaying scientific learning of the flora, fauna and resources of the island, as well as its inhabitants, in order to encourage backers to provide more money and persuade new colonists to undertake the dangerous voyage to the Americas. De Bry's magnificent edition appears to stretch the same text towards a wider series of functions. As well as accurately reproducing raw data and promoting the colonies, it also celebrates English success in the Americas and, most significant of all, examines the importance of the discovery of the new continent. De Bry's whole project has a strongly anti-Spanish agenda – hardly surprising given his own views, nationality, narrow escapes from persecution, and age at the time of the Saint Bartholomew's Day massacre (early forties) – and his framing of Harriot's narrative is explicitly theological.[19] Put simply, the iconography of *America* shows that Protestant colonialism has the potential to establish peaceful cooperation with the native Americans, leading to their conversion to the true faith; Spanish colonialism can lead only to dark visions of hell. De Bry generally represents compliant natives alongside Protestants and ferocious, savage natives resisting Spanish rule; a notable example being an engraving accompanying Girolamo Benzoni's account of his travels in South America, where the Darien Indians pour molten gold down the throat of a bound Spanish soldier, taunting the dying man with the words, 'Eat, eat gold, Christian', a traditional punishment carried out by devils in hell for the deadly sin of greed.[20]

White's drawings have generally been accepted as a faithful record of the now vanished Algonkian Indians of Virginia.[21] De Bry is, as a rule, reasonably accurate in his reproduction of White's originals as engravings, although he does have a tendency to 'idealize the features and soften the more awkward gestures', as well as adding important details to the last six drawings of the Pictes which 'in the olde tyme dyd habite one part of the great Bretainne'.[22] The former alterations may have been as much the result of the problems of reproducing drawings as bookplates and fitting them into an acceptable and recognisable style, as they were consciously motivated and significant changes. Equally, some backgrounds may have been reused from other texts as an economy measure and may not have the ideological force that we may want to attribute to them. However, it is clear that there is far more than questions of accuracy at stake in these elaborate and sophisticated representations of the native Americans, as the first engraving makes clear.

This prefatory engraving, attributed to De Bry himself, if analysed in terms of the subsequent sequence of images, appears to suggest that readers should interpret Harriot's text in an allegorical or typological manner. It shows a naked Adam and Eve immediately prior to the Fall, with the distinctly humanoid serpent entwined around the tree between them pointing upwards to the apples. Adam leans on the tree and looks skywards as if in contemplation and doubt about what is to happen, while Eve has an apple in her hand, which she is about to pick from the tree. In the foreground a lion lies down next to a mouse; a panther prowls on the side of the tree just behind Adam, while a mouse sits just behind Eve – all evidently symbolising the harmony of nature before the Fall. The background is divided into two sections by the tree; on the left, a mother nurses a baby; on the right, a man tills the ground with a staff, indicating the division of the sexes after the Fall and God's injunction in Genesis 3:19 that 'In the sweat of thy face shalt thou eat thy bread, till thou return unto the ground.'

The engraving makes a clear link between the Fall and the discovery of the Americas, especially as the first engraving in the sequence proper shows *The arrival of the Englishmen in Virginia*. But how the comparison is to be read is deliberately left ambivalent. Have the English rediscovered the Garden of Eden in the Americas? The Algonkians are represented positively throughout the sequence, shown to have sensible rules and regulations enabling them to produce an ordered society; to have enough knowledge of agricultural practices to produce an abundance of crops (see the background to Figure 12.1); to be skilled in the arts of technology and manufacturing (for example, 'Their manner of makinge their Boates'); to have a developed sense of religious practices; and to be able to look after their children well and be able to form other emotionally- and socially-stable relationships. They live in a desirable rural world, free from many of the ugly excesses of civilised life. The captions to Plates XV ('Their seethynge of their meate in earthen pottes') and XVI ('Their sitting at meate') contain explicit criticisms of English excess, as well as showing healthy, athletic natives performing these ordinary tasks. The first concludes with the admonishment, 'For wee should bee free from many kynes of diseases which wee fall into by sumptwous and unreasonable banketts, continuallye devisinge new sawces, and provocation of gluttonnye to satisfie our unsatiable appetite'; the second, with the observation, 'They are verye sober in their eatinge, and trinkinge, and consequentlye verye longe lived because they doe not oppress nature'.[23] Algonkian life – and, by an implicitly logical extension, the life new colonists would be able to enjoy in the Americas – provides mankind with all the benefits of the simple rural life advocated in biblical and classical literature, even if it is georgic rather than pastoral.[24] Or is the English presence in the Americas a violation, a forced entry into an earthly paradise which will lead to its imminent destruction? Clearly this is how the reader is asked to read the Spanish, rather than the English, presence in the Americas in De Bry's collection. However, Harriot admits that English behaviour in Roanoke has not always been as exemplary as it should have been: 'some of our companie towardes the end of the yeare, shewed themselves too fierce, in slaying some of the people, in some towns, upon causes that on our part, might easily have been borne withall'.[25] Here, the distinction between Protestant English and Catholic Spanish cannot easily be maintained. Such questions are posed at the start of the sequence of pictures and so provide an interpretative framework for subsequent images, as well as Harriot's text.

The engravings and their commentaries establish Virginia as a land of plenty which is administered in a civil and sophisticated manner by the natives.[26] Plate IX (Figure 12.1) shows 'An ageed manne in his winter garment', standing in front of a well-ordered, fortified village. To either side are fields of abundant corn and in the background, a line of carefully planted and pruned trees separates the village from the river on which the Indians paddle canoes. The accompanying caption describes the garment of the old man, but concludes with an acknowledgement that the surrounding details bear considerable significance: 'The contrye abowt this plase is soe fruit full and good, that England is not to be compared to yt', again suggesting that the simple life of the native Americans is superior to that experienced by many in England, as well as providing straightforward propaganda for the colonial enterprise.[27] The caption also reminds us how carefully we need to read the details in each engraving.

However, Plate XIII, 'Their manner of fishynge in Virginia' (Figure 12.2), provides a less optimistic assessment of the Indians. The illustration shows a canoe in the foreground with two Indians using nets to catch the abundant supply of various fish in the water, a number in the background wading in the water using spears and a series of large nets designed to force the fish nearer to their doom. Although Harriot's commentary recognises the skill of the Algonkians ('Ther was never seene amonge us soe cunninge a way to take fish withall'), the conclusion reveals that the question of religion is an ever-present problem:

> Dowbteles yt is a pleasant sighte to see the people, somtymes wadinge and

An ageed manne in his winter IX.
garment.

THe aged men of Pommeioocke are couered with a large ſkinne which is tyed vppon their ſhoulders on one ſide and hangeth downe beneath their knees wearinge their other arme naked out of the ſkinne, that they maye bee at more libertie. Thoſe ſkynnes are Dreſſed with the hair on, and lyned with other furred ſkinnes. The yonnge men ſuffer noe hairr at all to growe vppon their faces but aſſoone as they growe they put them away, but when thy are come to yeeres they ſuffer them to growe although to ſay truthe they come opp verye thinne. They alſo weare their haire bownde op behynde, and, haue a creſte on their heads like the others. The contrye abowt this plaſe is ſoe fruit full and good, that England is not to bee compared to yt.

B

12.1 'An ageed manne in his winter garment' in Thomas Harriot, *A Briefe and True Report of the New Found Land of Virginia* (1590), 9.

XIII.

Their manner of fishynge in Virginia.

THey haue likewise a notable way to catche fishe in their Riuers. for whear as they lacke both yron, and steele, they faste vnto their Reedes or longe Rodds, the hollowe tayle of a certaine fishe like to a sea crabb in steede of a poynte, wehr with by nighte or day they stricke fishes, and take them opp into their boates. They also know how to vse the prickles, and pricks of other fishes. They also make weares, with settinge opp reedes or twigges in the water, which they soe plant one within a nother, that they growe still narrower, and narrower, as appeareth by this figure. Ther was neuer seene amonge vs soe cunninge a way to take fish withall, wherof sondrie sortes as they fownde in their Riuers vnlike vnto ours. which are also of a verye good taste. Dowbtless yt is a pleasant sighte to see the people, somtymes wadinge, and goinge somtymes sailinge in those Riuers, which are shallowe and not deepe, free from all care of heapinge opp Riches for their posterite, content with their state, and liuinge frendlye together of those thinges which god of his bountye hath giuen vnto them, yet without giuinge hym any thankes according to his desarte. So sauage is this people, and depriued of the true knowledge of god. For they haue none other then is mentionned before in this worke.

12.2 'Their manner of fishynge in Virginia', in Thomas Harriot, *A Briefe and True Report of the New Found Land of Virginia* (1590), XIII.

> goinge somtymes in those Rivers, which are shallowe and not deepe, free and livinge frendlye together of those thinges which god of his bountye hath given unto them, yet without givinge hym any thankes accordinge to his desarte. So savage is this people, and deprived of the true knowledge of God. For they have none other than is mentioned before in this worke.[28]

The reader is referred back to Harriot's lengthy discussion of Algonkian religion, which he judges to be 'farre from the truth'.[29] Harriot describes how the Algonkians worship a number of Gods; nevertheless, they believe that one superior God made the world, then the other, lesser gods and goddesses, and finally, the sun, moon and stars. They believe that women were created before men and that the soul is immortal. Religion occupies a central role in their society and is administered by priests known as *wiroances*. Harriot argues that the Algonkians are 'not so sure grounded, nor give such credite to their traditions and stories but through conversing with us they were brought into great doubts of their owne, and no small admiration of ours'.[30] In short, they are ripe for conversion to Christianity, a process that will take place soon.

Harriot's observations, read alongside the commentary to Plate XIII, reverse and complicate the assumption of American superiority to English society and culture asserted in other illustrations. There the desirable flow of cultural traffic is the other way round and it is the Algonkians who must learn from the colonisers. In terms of the illustration of Adam and Eve, the Algonkians now appear as flawed, sinful, postlapsarian creatures, frozen in time after Adam and Eve have eaten the apple, not before, thus suggesting that the picture performs more work than simply representing the Indians as noble savages. In terms of the debates carried out in Spain referred to above, the Algonkians are represented as *both* pre-and postlapsarian men and women, indicating that Harriot's *Report*, alongside its manifold other functions, asks the reader to consider the problems, possibilities and wonder of the New World, and think where the natives fit into the grand scheme of things. Text and images combine to confront the English reader with fundamental ethnological and theological questions about the significance of the discovery of the New World.

This interrogative process is taken a stage further still in the final collection of illustrations: 'Som Picture, Of The Pictes which in the olde tyme dyd habite one part of the great Bretainne'. These are advertised as by 'The painter of whom I have had the first Inhabitans of Virginia' (John White), found in an 'oolld English chronicle … for to showe how that the Inhabitants of the great Bretannie have bin in times past as savuage as those of Virginia'.[31] The engravings show three pictures of Picts; a naked warlike man, covered in body-painting, holding a spear and the severed head of a vanquished enemy in his right hand, with other severed heads beside his feet in the foreground; a Pictish woman, similarly naked and body-painted with stars and moons, carrying three spears; and the daughter of the Picts, legs crossed at the calf, reclining on a spear, hair flowing freely behind her, staring to her left in a wary and aggressive manner.[32] All three carry large swords hanging from belts on their left sides. The last two pictures are of figures described as 'neighbour[s] unto the Picte'. They are not positively identified but they appear to be more civilised than the Picts (see Figure 12.3, 'The truue picture of a woman nigbour to the Pictes'): they wear clothes which cover their genitals (unlike the naked Adam and Eve); the man does not retain the severed heads of his enemies; and they look out to the side of the engraving rather than staring directly at the observer in a hostile manner.

The background details of all these pictures may – or may not – possess significance. The Picts appear to have turned their backs on the civilisation which is represented in the background to their pictures; the Pictish man has no obvious connection to the village or the ships which sail in the river behind him; the Pictish woman appears to be excluded from the castles on the hills behind her, which may have been erected to keep her out; and the daughter of the Picts stands in striking contrast to the orderly village behind her, her freely-flowing hair a sign of her savage wildness. In contrast, the neighbours of the Picts are integrated into their backgrounds: the three figures in the background to the man wear the same clothes as the main figure, indicating that they probably all come from the same village on the far shore and have built the large ocean-going boats (significantly larger than in the first two pictures) sailing away on the sea (towards new lands?); the woman in Figure 12.3 is also obviously from the village behind her, a settlement which may be intended to recall the shelter in the prefatory engraving of Adam and Eve.

What is the point of the inclusion of these engravings? Who is being compared to what, and why? Is this an example of early scientific ethnography? Do the pictures expose the ideological assumptions of the collaborators who produced this composite text – De Bry, Hakluyt, Harriot and, possibly, Raleigh – or are they more cunningly designed? Perhaps the first clue to the purpose of these plates is the deliberate omission of the name of the neighbours of the Picts, especially given that they are clearly identified as the Britons in John White's drawings. The iconography of the engravings indicates that one group of peoples, the Picts, are savages in need of civilising; the other group, the Britons, have established the rudiments of agricultural life and look out towards the New World, perhaps ready to colonise and civilise others, if not now, then in the future.[33] Any reader of Richard Hakluyt's *Principall Navigations* would have encountered not only Geoffrey of Monmouth's account of the early Arthurian British empire in Europe, but the Welsh prince Madoc's discovery of Mexico and establishment of colonies later proved by (rather dubious) linguistic evidence.[34]

De Bry's refusal to name the Britons is unsettling in itself, given the savage

12.3 'The truue picture of a woman nigbour to the Pictes', in Thomas Harriot, *A Briefe and True Report of the New Found Land of Virginia* (1590), V.

reputation of the ancient Picts as the enemies of the Britons, who are here represented as if they were the dominant people who inhabited the ancient island of Britain.[35] Britain is therefore depicted as a savage land in contrast to the relatively civilised society of the Algonkians, who resemble the Britons far more closely than they do the Picts. However, other details suggest that the opposite comparison is the one that should be made and that the Algonkians should be compared to the Picts. Much is made of the body-painting of both peoples, a practice which provides a neat link between the two sequences of pictures. Plate XXIII, 'The Marckes of Sundrye of the Cheif mene of Virginia' (Figure 12.4), the last one in the series of Indian life, shows the tattoos used by the Virginians – marks inscribed on the body which link them to the Picts rather than the Britons. This point is reinforced by the other negative connotations established by the other closing images which emphasise the Algonkians' paganism. Read this way, the engravings would seem to suggest that the English in the late sixteenth century are the – literal and spiritual – descendants of the Britons, who have to build a new Britain in the Americas by conquering and absorbing the natives.[36]

Once again, the *Report* proves to be a studious exercise in balanced ambivalence. The Virginians are in need of the Christian religion of the English, but they have much to teach their future colonisers about sensible bucolic living. In some ways they are savages (Picts); in others, they are civilised (Britons). The work seems to remind the reader that the future is not written in stone and the possibilities for the 'New Britain' in the Americas can go in more than one direction. The world is poised at a similar juncture to that represented in the first engraving of Adam and Eve with the apple weighed in Eve's hand but still attached to the tree. Britain's past may well be a guide to the future; alternatively, the future may point back even further to the harmony experienced by mankind at the world's beginning in the Garden of Eden. The civilised but pagan inhabitants of the Americas may even help to explode the old established dichotomies between civil and savage, leading to a future society beyond the brutalities of the old world with its vicious sectarian conflicts.[37] De Bry's work represents the discovery of the Americas as an apocalyptic event, the discovery of new wonders that cannot be quantified or classified within the realms of Western scientific or ethnological knowledge, exactly as the parallel debates in Spain suggested.[38] The content has gone beyond the form, and the identities of the coloniser are called into question as much as those of the colonised.[39] For, if the Algonkians are more like the Britons than the Picts, then what relationship do they bear to the English? Are all such men brothers?

De Bry's framing of Harriot's *Report* can – arguably – be seen as very much in line with the text (hardly surprising, if the two men were collaborators). The third part of Harriot's work describes the 'nature and

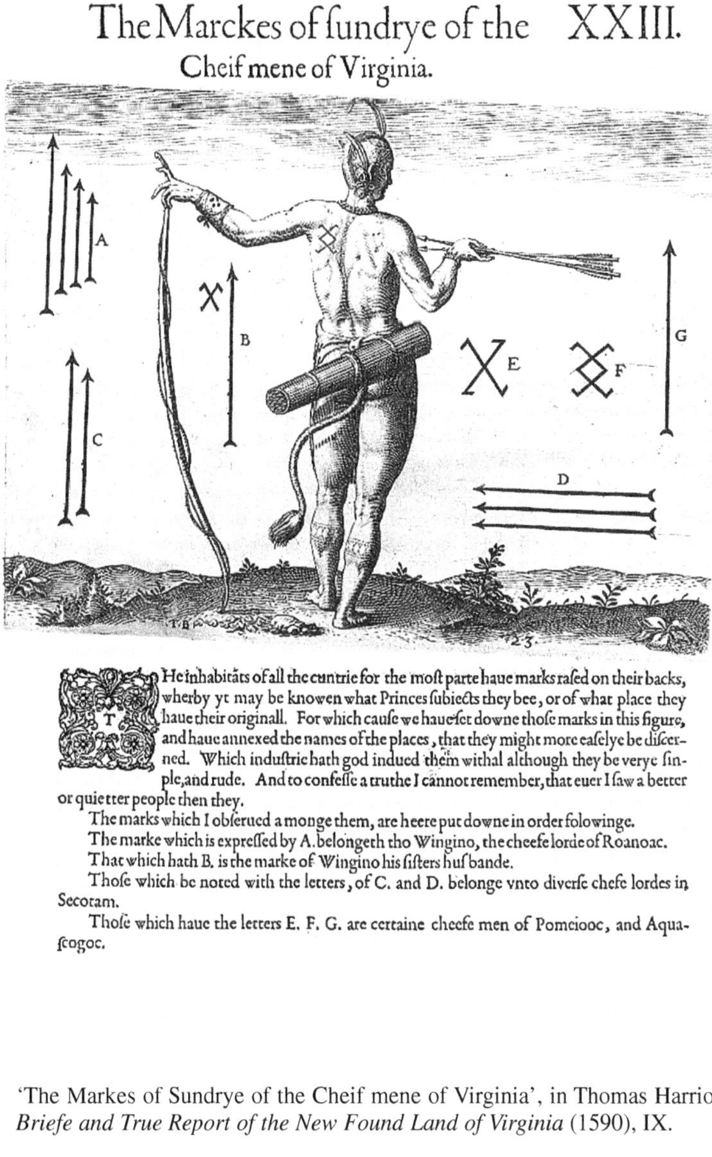
The Marckes of ſundrye of the XXIII.
Cheif mene of Virginia.

THe inhabitats of all the cuntrie for the moſt parte haue marks raſed on their backs, wherby yt may be knowen what Princes ſubiects they bee, or of what place they haue their originall. For which cauſe we haue ſet downe thoſe marks in this figure, and haue annexed the names of the places, that they might more eaſelye be diſcerned. Which induſtrie hath god indued them withal although they be verye ſinple, and rude. And to confeſſe a truthe I cannot remember, that euer I ſaw a better or quietter people then they.

The marks which I obſerued amonge them, are heere put downe in order folowinge.

The marke which is expreſſed by A. belongeth tho Wingino, the cheefe lorde of Roanoac.

That which hath B. is the marke of Wingino his ſiſters huſbande.

Thoſe which be noted with the letters, of C. and D. belonge vnto diverſe chefe lordes in Secotam.

Thoſe which haue the letters E. F. G. are certaine cheefe men of Pomeiooc, and Aquaſcogoc.

12.4 'The Markes of Sundrye of the Cheif mene of Virginia', in Thomas Harriot, *A Briefe and True Report of the New Found Land of Virginia* (1590), IX.

manners of the people', registering a similar awareness to that of the sequence of engravings. Harriot notes that the Virginians are not at all sure of their religion and are open to persuasion.[40] This open-mindedness leads to a moment of great embarrassment after Harriot has explained the contents of the Bible to them. Their excessive zeal for the truth makes them profane the book as a sexual rather than a religious object as if somatic sensation can replace intellectual effort:

> And although I told them the booke materially & of it self was not of anie such vertue, as I thought they did conceive, but onely the doctrine therein contained; yet would many be glad to touch it, to embrace it, to kiss it, to hold it to their brests and heades, and stroke over all their bodie with it; to shewe their hungrie desire of that knowledge which was spoken of.[41]

While the first trait illustrates a tolerance which contrasts favourably with the brutal zeal of internecine European politics, the second opens out the problem once again, representing the Algonkians as benign, confused and inferior people in need of the guidance of their cultural superiors. One moment the Indians are savage critics, the next, they are helpless savages.[42] The gap between the two peoples extends and contracts in a series of rapid movements.

However, the *Report* ends with a phenomenon that no one can explain, the death of hostile Indians after the visits of the English.[43] In fact, events would seem to confirm the simultaneous bemusement of both English and Algonkians alongside an efficacious serendipity, leaving the text balanced between success and failure – the actual position of English colonial ventures in the period. The Algonkians are convinced that the English can kill whom they please without weapons and ask them to help destroy their enemies. Although the English refuse, arguing that such entreaties are ungodly and that all should try and live together and wait for God to act, what the natives want is what actually happens, leaving the colonisers in a more powerful position than before and allowing them to retain the moral high ground:

> Yet because the effect fell out so sodainly and shortly after according to their desires, they thought nevertheless it came to passe by our meanes, and that we in using such speeches unto them did but dissemble the matter, and therefore came unto us to give thankes in their manner that although wee satisfied them not in promise, yet in deedes and effect we had fulfilled their desires.[44]

Although the two peoples misunderstand each other, events conspire to leave both satisfied; the Algonkians get rid of their enemies and the English confirm their colonial power. Nevertheless, the gratitude of the natives depends upon their perception of the English as duplicitous, hinting that future discord may result and, leaving matters, yet again, in a state of precarious balance.[45]

Perhaps the most important point for the European reader is that the text does not conclude with a discussion of the identity of the newly discovered peoples, but the identity of the English colonisers, forging a powerful link with the illustrations of the ancient inhabitants of Britain. The final pages of Harriot's *Report* list the various opinions the Algonkians have of the nature of the English; some suggest that they are not born of women but are immortals; some prophesy that more English will come to kill them in future years; others believe that these creatures are ethereal and shoot invisible bullets at the hapless natives who anger them. It is the Indians who ask the questions, not the English; a surprising reversal of expected norms, indicating that the new knowledge which had brought them back from the Americas tended to undermine expected and established norms of enquiry and would require urgent and strenuous rethinking before the wonders of the New World could be truly understood.

A Briefe and True Report of the New Found Land of Virginia was clearly intended as a work that would contribute to the scientific understanding of the Americas; at the same time it had the openly advertised design of encouraging colonial ventures to the Americas, being addressed 'To the Adventurers, Favourers, and Wellwillers of the Enterprise for the inhabiting and planting of Virginia'.[46] Moreover, as the first illustration of Adam and Eve and the conclusion of the *Report* indicate, scientific questions could not be considered without reference to religious issues, an ethnology that was also theological.[47] And this turn to religion also meant that for Harriot and De Bry an English colonial empire in North America could not be achieved unless the English rethought their own identities and historical genealogy in order to meet the challenge of the New World. Scientific and historical enquiry would have to face inwards as well as outwards.

Notes

1. Hogden, M.T. (1964), *Early Anthropology in the Sixteenth and Seventeenth Centuries*, Philadelphia: Philadelphia University Press, p. 1.
2. Alington, G. (1996), *The Hereford Mappa Mundi: A Medieval View of the World*, Leominster: Fowler Wright Books.
3. Hulme, P. (1986), *Colonial Encounters: Europe and the Native Caribbean, 1492–1797*, London: Methuen, pp. 16–43.
4. Bodin, J. ([1566] 1945), *Method for the Easy Comprehension of History*, trans. B. Reynolds, New York: Columbia University Press; Hogden (1964), pp. 113–14.
5. Greenblatt, S. (1991), *Marvelous Possessions: The Wonder of the New World*, Oxford: Clarendon Press, p. 14. See also Linton, J.P. (1998), *The Romance of the New World: Gender and the Literary Formations of English Colonialism*, Cambridge: Cambridge University Press. Linton examines how the new and unfamiliar experience of the Americas was analysed through the use of romance narratives.

6. Pagden, A. (1982), *The Fall of Natural Man: The American Indian and the Origins of Comparative Ethnology*, Cambridge: Cambridge University Press, pp. 27–55.
7. Ibid., p. 38.
8. Ibid., ch. 4.
9. Ibid., chs 5–6. See also Pagden, A. (1993), *European Encounters with the New World*, New Haven: Yale University Press, pp. 56–7.
10. Hogden (1964), pp. 121–8; Pagden (1982), introduction.
11. For one very influential attempt to explain this phenomenon, see Foucault, M. (1970), *The Order of Things: An Archaeology of the Human Sciences*, trans. from the French, London: Tavistock, pp. 17–45.
12. Kidd, C. (1999), *British Identities Before Nationalism: Ethnicity and Nationhood in the Atlantic World, 1600–1800*, Cambridge: Cambridge University Press, p. 10.
13. For recent analyses of the *Report*, see Greenblatt, S. (1988), 'Invisible Bullets', in *Shakespearean Negotiations: The Circulation of Social Energy in Renaissance England*, Oxford: Clarendon Press, pp. 21–65; Sokol, B.J. (1994), 'The Problem of assessing Thomas Harriot's *A Briefe and True Report of the New Found Land of Virginia*', *Annals of Science*, **51**, 1–16; Albanese, D. (1996), *New Science, New World*, Durham, NC: Duke University Press, pp. 24–9. While I have learnt much from all three essays, I would suggest that Greenblatt's argument that Harriot's text exposes the 'function of illusion in the establishment of religion' (p. 39); Sokol's that Harriot was one of a number of 'genuine explorers who could question or transcend embedded schemes of valuation' (p. 16); and Albanese's that the English and the British are the same people and that the Picts are 'the precursors of the English race' (p. 24), all limit their respective findings.
14. Quinn, D.B. (ed.) (1955), *The Roanoke Voyages, 1584–1590*, 2 vols, London: Hakluyt Society, vol. 1, pp. 38 and 314; Quinn, D.B. (ed.) (1974), *The Hakluyt Handbook*, 2 vols, London: Hakluyt Society, vol. 2, p. 376.
15. For further analysis, see Hadfield, A. (1998), *Literature, Travel and Colonial Writing in the English Renaissance, 1545–1625*, Oxford: Clarendon Press, pp. 99–100 and 112–26.
16. Quinn (1955), vol. 1, p. 39. See also Hulton, P. (ed.) (1984), *America 1585: The Complete Drawings of John White*, London: British Museum, pp. 17–21.
17. For details, see Kupperman, K.O. (1984), *Roanoke: The Abandoned Colony*, Savage, MD: Rowman & Littlefield, pp. 106–21. For an account of John White's life and importance, see Hulton, P. (1978), 'Images of the New World: Jacques Le Moyne de Morgues and John White', in K.R. Andrews, N.P. Canny and P.E.H. Hair (eds), *The Westward Enterprise: English Activities in Ireland, the Atlantic and America, 1480–1650*, Liverpool: Liverpool University Press, pp. 195–214.
18. De Bry published other works which demanded the reader have an ability to read iconographically, and he was clearly well-versed in the modes of Protestant allegory; see, for example, the emblem book, *Iani Iacobi Boissardi Vesvntini Emblematum* (Frankfurt, 1593).
19. Something of an irony given Harriot's scandalous reputation in England; see Greenblatt (1988), pp. 22–3; Shirley, J.W. (1974), 'Sir Walter Raleigh and Thomas Harriot', in Shirley, J.W. (ed.), *Thomas Harriot: Renaissance Scientist*, Oxford: Clarendon Press, pp. 16–35 and 23–4.
20. Alexander, M. (ed.) (1976), *Discovering the New World, Based on the Works of Theodor De Bry*, London: London Editions, p. 137.
21. See Hulton, P. and Quinn, D.B. (eds) (1964), *The American Drawings of John*

White, 1577–1590, 2 vols, London: British Museum, for the most comprehensive analysis. See also Sokol (1994).

22. Paul Hulton, ed., introduction to Harriot, T. (1972), *A Briefe and True Report of the New Found Land of Virginia*, New York: Dover, p. xi.
23. Harriot (1972), pp. 60–61.
24. See Rivers, I. (1979), *Classical and Christian Ideas in English Renaissance Poetry*, London: George Allen & Unwin, pp. 9–20.
25. Harriot (1972), p. 30. This is probably a criticism directed at Ralph Lane's attempts to run the colony along military lines; Quinn (1955), vol. 2, p. 381.
26. For commentary, see Kuppermann, K.O. (1980), *Settling with the Indians: The meeting of English and Indian cultures in America, 1580–1640*, Totowa, NJ: Rowman & Allanheld.
27. Harriot (1972), p. 52.
28. Ibid., p. 56. For other disparaging comments on the Algonkians' lack of religion, which is highlighted as the sequence develops, see Plate XXI, 'Ther Idol Kivvasa', where Harriot condemns their idolatry and comments, 'Thes poore soules have none other knowledge of god although I thinke them verye Desirous to know the truth' (p. 71). This plate is probably based on the experiences of Le Moyne in Florida rather than those of White and Harriot in Virginia, and may have been introduced by De Bry to help conclude the sequence of illustrations with a sustained attack on the idolatry of the Algonkians, emphasising their need for conversion. See Hulton, P. (ed.) (1977), *The Workes of Jacques Le Moyne de Morgues: A Huguenot Artist in France, Florida and England*, 2 vols, London: British Museum, p. 216; Hulton and Quinn (1964), vol. 1, pp. 93–4. See also Plate XXII, 'The Tombe of their Werowans or Cheiff Lordes' (Harriot, 1972, pp. 72–3).
29. Ibid., p. 25.
30. Ibid., p. 27.
31. Ibid., p. 75. Plate III, 'The truue picture of a yonge dowgter of the Pictes', is actually based on a painting by Jacques Le Moyne de Morgues (Hulton, 1977, vol. 2, Plate 7). De Bry may have confused the collections he had acquired from the two artists and the remaining four pictures may well be based on lost work by Le Moyne (Hulton, 1978, p. 211).
32. Hulton argues that this last detail, a change from the original Le Moyne drawing, in which she looks straight ahead, has been made to 'give it [her face] a more classical appearance' (1978, p. 211). It might also be argued that the change makes the daughter look more threatening to the reader of the text.
33. This is what the Britons were reputed to have done.
34. Hakluyt, R. ([1600] 1903), *The Principal Navigations, Voyages, Traffiques & Discoveries of the English Nation*, 12 vols, Glasgow: MacLehose, vol. 5, pp. 133–5; Geoffrey of Monmouth (1966), *The History of the Kings of Britain*, trans. L. Thorpe, Harmondsworth: Penguin, p. 7.
35. See Hadfield (1998), pp. 119–22.
36. The concept of a New Britain is implicit in the colony's name 'Virginia', chosen in honour of Elizabeth I; later treatises such as William Strachey's *The Historie of Travell into Virginia Britania* ([1612] 1953), ed. L.B. Wright and V. Freund, London: Hakluyt Society, made the comparison explicit; Strachey refers to Virginia as 'Nova Britania' (p. 10).
37. See Hadfield (1998), ch. 2. A similar argument is made by Montaigne ([1603] 1910), 'Of the Cannibals', in *Essayes*, trans. J. Florio, 3 vols, London: Everyman, vol. 1, pp. 215–29.

38. See Greenblatt (1991), introduction.
39. On this frequently discussed question in post-colonial theory, see Bhabha, H. (1994), 'Articulating the Archaic: Cultural difference and colonial nonsense', in Bhabha H., *The Location of Culture*, London: Routledge, pp. 123–38; Memmi, A. (1965), *The Coloniser and the Colonised*, trans. H. Greenfield, New York: Orion, conclusion.
40. Harriot (1972), pp. 24–30.
41. Ibid., p. 27.
42. On the 'savage critic', see Pagden, A. (1983), 'The Savage Critic: Some European Images of the Primitive', *Yearbook of English Studies*, **13**, 32–45.
43. The most searching analysis of this problem is Greenblatt (1988), pp. 24–39.
44. Harriot (1972), p. 29.
45. Compare the story of the death of Captain Cook; see Rennie, N. (1995), *Far-Fetched Facts: The Literature of Travel and the Idea of the South Seas*, Oxford: Clarendon Press, pp. 129–36.
46. See Hadfield, A. (1991), 'Writing the New World: More "Invisible Bullets"', *Literature and History*, **2** (2), 3–19 (pp. 9–10).
47. For related observations of early-modern science, see Debus, A.C. (1978), *Man and Nature in the Renaissance*, Cambridge: Cambridge University Press, *passim*.

CHAPTER THIRTEEN

'Adding to the World': Colonial Adventure and Anxiety in the Writings of John Donne

Richard Sugg

In 'A Valediction: Of Weeping' the speaker, presenting the two lovers as a sufficient world in themselves, likens the reflection of his mistress's image in his teardrop to the process of map-making:

> On a round ball
> A workman that hath copies by, can lay
> An Europe, Afric, and an Asia,
> And quickly make that, which was nothing, all.[1]

The curious omission of the American continents from the world of this poem is far from accidental. Before and during Donne's lifetime, the 'Indies' of both north and south America were in many ways so alien to the mentality of European Christendom that in certain crucial senses they indeed did not fully exist.[2] This shadowy, liminal status was not, however, simply a result of European obliviousness, uninterest or contempt. The startlingly alien inhabitants, culture and natural resources of America were if anything too strange and potentially overwhelming to be absorbed easily by Christian cosmology. At the same time, the profound otherness of the new world offered certain intoxicating, half-guessed possibilities – both actual and psychological – whose very danger was at times irresistibly attractive. In considering Donne's own oscillation between colonial excitement and colonial anxiety, and his success in at least provisionally fusing these opposed responses, I will focus on the evangelical aims of his Virginia Sermon of 1622. It is interesting to note, however, that Donne's attitudes to America retained an underlying continuity, though poured into different moulds in different historical and biographical circumstances. His secular poetry and sacred prose are united by the need for a psychological energy which, equalling that of the new world in its vigour and intensity, is able to exploit and creatively transform the fearful wonder of the Indies.

Also common to poems and sermons is a recourse to Aristotelian beliefs about the status of heathen lands and peoples. Aristotelian arguments were extremely pervasive in early-modern literary representations, and in the rationalising and justification of colonial conquest. Two particular strands of Aristotle's thought proved especially durable. The doctrine of natural slavery held, firstly, that certain 'uncivilised' peoples of the world were naturally, intrinsically slaves to their superior masters.[3] In 1550–51 Juan de Sepúlveda (1490–1573), the Aristotelian scholar and historiographer of Emperor Charles V, had used Aristotle's theories in a notorious defence of Spanish colonial atrocities, during the debate against Bartolomé de las Casas (1474–1566), the period's most famous European campaigner for Indian rights.[4] Elizabethan and Jacobean responses to the Spanish precedent are complex. Las Casas' *Breve relación de la destrucción de las Indias Occidentales* (Seville, 1552) was first published in English in 1583, and a gruesomely illustrated Latin version appeared in Europe in 1614.[5] Whatever its precise factual basis, the 'Black Legend' of Spanish cruelties was certainly well-established and well-credited within England by the early seventeenth century.[6] Nonetheless, many English writers continued to employ casual or systematic versions of Aristotle's arguments in colonial tracts.

A second factor not always sufficiently emphasised in accounts of Aristotelian colonial legitimising is the crucial role of Renaissance pneumatology. In the *Politics* the fundamental inferiority of the natural slave is quickly associated with the hierarchies of soul and body, and of male and female, which ramify so extensively through early-modern colonial thought.[7] Notwithstanding increasingly critical attitudes toward Aristotle, such ideas clearly remained pervasive at this point. The still inextricable questions of generation and pneumatology were both rooted in the Aristotelian notion of 'a body and material mass' drawn 'from the female', and a soul, 'the substance of a particular body … which fashions it' from the male.[8] No less important than this sense of formless, feminine matter *in potentia* – which Aristotle himself analogises with 'soil' – is the implication that such merely raw material does not quite properly exist. For Aristotle an in-forming soul is an *a priori* condition of 'a natural organised body'.[9] As Margarita Zamora has shown, the 'monstrous femininity' of America had a definitely Aristotelian logic, monstrosity merely lying further along a continuum of incompleteness, and excessively predominant female matter.[10]

Donne himself clearly adhered to a basically Aristotelian pneumatology.[11] Evident in the very flippancy of the 'Problem', 'Why hath the common opinion affoorded woemen Soules?' and in the observation that 'Man to God's image, Eve, to man's was made, / Nor find we that God breathed a soul in her', the flawed pneumatology of woman reveals its Aristotelian basis most clearly in a sermon of 1622.[12]

> [Although] we are sure *women* have Soules as well as *Men*, ... yet it is not so expressed, *that God breathed a Soule* into *Woman*, as hee did into Man; All formes of Government have this Soule, but yet God infuseth it more manifestly, and more effectually ... in a *Kingdome* .... [Consequently,] this form of a *Monarchy*, of a *Kingdome*, is a more lively, and a more *masculin* Organe, and Instrument of this Soul of Soveraigntie, then the other forms are.[13]

This parallel relies on the notion that women's relative inactivity is not a casual fondness for ease, but a necessary result of their weaker, less thorough pneumatological composition. Women are, indeed, not merely inactive, but arguably 'in-actual' by comparison with men. Because of the close mutual interaction of the Renaissance body and soul, physical, spiritual and intellectual passivity are all inextricably associated. Although women may occasionally exhibit 'mild innocence', 'active good' in them is 'a miracle' – a condition which mirrors the more general entanglement of the rational male soul and rational Protestant government.[14] Just as an 'idle body, is a disease in a State' so 'an idle soul, is a monster in a man...'. Significantly, this 'monstrosity' very precisely reflects the predominance of female matter over the male, in-forming soul identified by Zamora: 'that soul that does not think, not consider, cannot be said to *actuate* (which is the proper operation of the soul) but to evaporate; not to work in the body, but to breathe, and smoak through the body'.[15] That individual who is morally required to govern himself by an active rationality is still more morally obliged to assist in the rational control and direction of intrinsically inferior beings – whether women or savages – who simply do not have such powers. Literally im-perfect or uncompleted, feminine America either cannot or must not exist on the same level as male, European Christendom. Donne himself, though relatively positive and enlightened in his colonial attitudes – and indeed ultimately too curious about America as a concrete, dynamic entity to subdue it within any watertight theoretical scheme – appears frequently to assume or depend on the blankness and malleability of the New World.

Donne's first known involvement with the revived plans for North American colonisation occurred in February 1608/9, when he attempted unsuccessfully to become secretary to the Virginia Company.[16] Despite his failure, he appears to have remained on friendly terms with his successful rival, William Strachey.[17] He also owned a copy of William Symonds's early Virginia Sermon, given on 25 April 1609, and the large number of his friends who were involved with the Company must have kept him informed of its often precarious fortunes in following years.[18] The colonial allusions of two verse letters which may date from 1609 or earlier, encapsulate the dualistic attitude found in Donne's Virginia Sermon. Writing to the Countess of Huntingdon, he refers to 'That unripe side of earth, that heavy clime / That gives us man up now, like Adam's time'. Despite the obvious primitivism of these New World inhabitants, and the fact that they 'bear the sin' of an Adam seemingly unknown to them, the image is relatively positive and optimistic.[19] The new, 'unripe' world is still potentially able to mature through stages of grace toward Christian salvation. A subtly altered version of America is seen in Donne's letter to the Countess of Bedford: 'We have added to the world Virginia, and sent / Two new stars lately to the firmament'.[20] America has been not merely discovered, but produced and perfected by European law, government, technology and religion – 'added' to the civilised, comprehensible world. The close association of Virginia with the recent 'new stars' suggests, however, an unconscious recognition that America had ruptured existing Aristotelian cosmology quite as drastically as the astronomical controversies of Donne's lifetime. If Virginia and America had been added, they had scarcely been assimilated.[21]

The most fundamental requirement for any such process was the location of the New World within accepted biblical history. From at least December 1621 to May 1623, Donne appears to have taken an increasingly close interest in the question of New World evangelism. On 22 May 1622, he had been made an honorary member of the Virginia Company, and on 3 July he was made both 'free of this Companie' and 'free of the Counsell'.[22] Shortly after his Virginia sermon of November 1622, he attended six (sometimes quite lengthy and fraught) meetings of the Company.[23] Between Christmas 1621 and spring 1623 Donne's sermons include at least six geographic, cartographic or explicitly American references or images, beyond those of the Virginia sermon itself,[24] three of which refer directly to overseas evangelism. His interest in Christianising the Virginians, then, although probably increased by his official contact with the Company, seems to have preceded his documented involvement with it. In December 1621 plans to systematically 'ground' and 'prepare' Indian minds were certainly afoot.[25] The longstanding project to establish a 'College' for the conversion of the Virginians, previously marginalised by more immediate practical concerns, at last looked a realistic possibility.[26] By November 1622, however, the prospects for Indian conversion had once again been seriously undermined; this time by the horrific events of 22 March, when the Indians – who by now were freely mingling with colonists in the latters' houses – suddenly and unexpectedly massacred 347 of the Virginia settlers.[27] Shortly after the news of this disaster reached London in early July the already barely-suppressed unwillingness to accept the Indians' existence exploded into demands for their outright extermination.[28] Amidst a mixture of explicit and precise schemes for violent conquest, newly-vigorous assertions of natural slavery, and general bestialisation, the impulse to genocide was nowhere more ferociously expressed than in *A Poem on the Late Massacre in Virginia* (London, 1622), by Donne's friend, Christopher Brooke.[29] Emphatically reinforcing the subhuman liminality of the Indian 'dregs' and 'garbage' who, like Donne's 'inconsiderate' man, are no more than 'Soules drown'd in flesh and blood', Brooke insists on a rightful 'Slaughter' which will leave 'not a Creature … [to] restore such shame of Men, and Nature'.[30]

How was Donne to respond to the challenge of delivering a positive, pacific text, and to represent evangelism as a realistic possibility, in these circumstances? In examining his sermon I will consider four aspects in particular. The first is how far Donne relies, explicitly and implicitly, on a version of the Aristotelian rhetoric employed by Brooke and others. The second, and perhaps most obvious, is an abstracting and spiritualising of the colony's past, present and future states; the third, a more indirect recognition of America as a concrete, novel, and ambiguous physical entity; and the fourth, an evident fusion of the New World's empirical reality with its projected scriptural role. I will argue that, in achieving this, Donne in fact incorporates and even emphasises the strangeness and liminality of America, rather than simply suppressing or evading it.

Aristotelian notions are implicitly present in Donne's paraphrase of the 'Law of *Nature* and of *Nations*', which declares any country rightfully open to colonisation 'if the inhabitants doe not in some measure fill the land' to 'bring foorth her increase for the use of men'.[31] This argument concerning *res nullius*, as Walter S. Lim has shown, hinged on a dual assertion of Indian agricultural neglect or inadequacy, and a typically Christian rationalisation which viewed settlement as piously maximising God's bounties, rather than simply exploiting his fellow creatures.[32] Donne's further requirement, for 'hammering and filing' of the inchoate new land, shows that his arguments cannot entirely be dissociated from those of Christopher Brooke.[33] At a quite basic level, moreover, the suggestion that the indigenous Virginians 'doe not in some measure fill the land' is philosophically complicit with Brooke's claim that their souls do not fill or define their bodies. Especially in the highly charged racial climate of late 1622, the implication that these uncivilised Indians could not form and structure their land was very liable to connote their similarly unregulated, unspirited bodies and minds.

The crucial difference, however, between the Aristotelianism of Donne and of Brooke is that Donne believes the Indians' souls to be perfectible through Christian grace. Some months before the Virginia sermon he had already presented the relatively formless Americans as positively open, in their spiritual blankness, to Christianity. Considering the case of 'a *Heathen man*, a meere naturall man, uncatechized, uninstructed in the rudiments of the Christian Religion', Donne imagines presenting him

> first with this necessitie; thou shall burn in fire and brimstone eternally, except thou believe a *Trinitie of Persons, in an unitie of one God*, Except thou believe the *Incarnation* of the second person of the Trinitie, the Sonne of God, Except

thou believe that *a Virgine had a Sonne*, and the same Sonne that God had, and that God was Man too, and being the immortall God, yet died.

The heathen's most likely response would, Donne admits, simply be 'not to believe *Hell* it selfe, and then nothing could binde him to believe the rest'.[34] A version of the argument from design, with its universal empirical appeal, would probably be more effective than the intellectual mysteries of Scripture.[35] Failing this, the actual Scriptures (rather than bare paraphrases) would 'have so orderly, so sweet, and so powerful a working upon the reason, and the understanding' that 'one who were altogether neutrall, disinteressed, unconcerned in either party, nothing towards a *Turke*, and as little toward a *Christian*' would necessarily favour the Christian Bible over the Muslim Koran.[36] The Indian has, then, a positive blankness which compares favourably with the delusions of Islam. Corresponding to what Lim calls *terrae nullius*, he is again 'unripe' – not beyond, but awaiting, Christian salvation.[37]

In keeping with this more optimistic, teleological attitude, Donne abstracts and rewrites the project, and America in general, in emphatically biblical terms. His chosen text is Acts 1:8: 'But yee shall receive power, after that the holy ghost is come upon you, and yee shall be witnesses unto me both in Jerusalem, and in all Judea, and in Samaria, and unto the uttermost part of the earth.' Heathen America is thereby scripturally located, albeit somewhat loosely, as this 'uttermost part of the earth', and the seeming impotence of the colonists' situation transcended, on the spiritual plane, by the promise of a different and more absolute 'power': the universal dissemination of Christian grace. Donne is consistently emphatic that the venture must be understood primarily in spiritual, not commercial terms. 'If you seeke to establish a temporall kingdome there, you are not rectified'; 'All that you would have by this Plantation, you shall not have; GOD bindes not himself to measures.'[38] The '*Patents* ... *Charters*' and '*Seales*' granted by James I must be understood as subordinate to 'your principall ende ... to gaine Soules to the glory of GOD' – an aim which 'authorises Authoritie, and gives power to strength it selfe'.[39] If the colonists 'would be as ready to hearken at the returne of a *Ship*, how many *Indians* were converted to *Christ Jesus*, as what Trees, or druggs, or Dyes that Shippe had brought, then you were in your right way, and not till then'.[40] In keeping with Donne's abstraction of the particular present into future spiritual glory, the massacre itself is absorbed almost seamlessly within the still greater, but crucially meaningful, calamity of the Deluge: 'though a Flood, a Flood of bloud have broken in upon them, be not discouraged'.[41]

The sermon's appeals for Indian conversion could be viewed as deliberate attempts to shift attention from the massacre and accompanying calls for retaliation. I want to argue, however, that Donne in fact felt personally compelled – rather than formally obliged – to advocate universal evangelism, and that this basic compulsion substantially influenced the design of his text. For all his careful integration of America into a meaningful Christian narrative, its location as the scriptural and actual 'uttermost part of the earth' raises an unavoidable question: 'Did the *Apostles* in person, preach the *Gospell*, over all the World?'[42] With characteristic thoroughness, Donne enumerates six distinct scriptural passages implying that Christianity should indeed have been '*preached to all the world*'.[43] Those Fathers of the Church, however, who supposed the Apostles to have performed 'an actuall and personall preaching ... over all the world' would, 'had they dream'd of this world ... discovered since ... never have doubted to have admitted a *Figure*, in that, *The Gospell was preached to all the world*'.[44] The 'West Indians' had in fact no more heard of the Christian faith than they had of Augustus' allegedly universal taxation.[45] 'All the world' signified all the known world, and Christ's injunction to preach his gospel universally, throughout the actual world, was yet to be fulfilled, referring to an apostolic '*Succession*'.[46]

Notwithstanding the ingenuity of his proposed response, to which I will shortly return, Donne has here touched on perhaps the most disturbing and finally unresolvable question raised by the New World. America, which had indeed been added to the world of European Christendom, seemed potentially extraneous to biblical history. This massive collision of the concrete and the abstract threw up a shower of difficult questions. Who were America's inhabitants? Had they, like the people of Europe, Asia, and Africa, really also descended from the sons of Noah? If so, why were they so unrecognisably savage and unChristian? Did they in fact represent the result of a second Creation, or were they the distant progeny of the lost tribes of Israel?[47] Was it possible that they had reached America from the Old World via an as yet uncharted neck of land?[48] Perhaps deeply suppressed within these and other debates was one central, unthinkable possibility: that Christ or God no more knew America, than it did them. Donne's sometime rival, William Strachey, had considered such questions as recently as 1612.[49] Strachey's own account, however, emphatically deferred to the authority of José de Acosta. A New World veteran of some 17 years, Acosta had disputed these and many other American conundra across 25 detailed and scrupulous chapters of his *Natural and Moral History of the Indies* (Seville, 1604).

Like Acosta, Donne was too intellectually curious and rigorous to be satisfied with glib or token answers. In recognising the figurative character of the New Testament 'world', the latter stresses – with a significant use of the present tense – how 'wee dispute with perplexitie, and intricacy enough, how any men came at first [to America], or how any beastes, especially such beastes as men were not likely to carry'.[50] It appears from this that Donne had almost certainly read Acosta, who expressly denies the possibility that 'in so long a voyage men would take the paynes to carrie Foxes to Peru …'.[51] Donne in fact refers to Acosta by name in the Virginia sermon, as giving 'very good reasons' against the performance of New World miracles, and he cites him as his authority for the same point on three other occasions.[52] Again, a few months before, in a sermon of 8 March 1621/22, Donne's discussion of the Antipodes looks strikingly like a compression of Acosta's argument.[53] The textual parallels of March and November collectively suggest not only that Donne read Acosta, but that he may well have been initially inspired to read – or reread – him just because of his increasing interest in the Virginian enterprise during 1622. Despite the ostensibly unworldly and scripturally-rooted quality of the Virginia sermon, its author appears in fact to have been minutely curious about an America whose strangeness and novelty were impressed on him by very dynamic topical pressures. One of his chief textual filters for this exotic realm – situated by his own admission beyond the supposedly absolute authority of scripture – was a book equally fascinated by the most precise empirical details of America's existence.

The impression that Donne found the mere existence of America unnerving is reinforced by a sermon delivered in September 1622, two months before the Virginia Company address. Speaking in support of James I's 'Directions for Preachers', Donne divides his chosen text (Judges 5:20) into the 'two *Hemispheres* of the world, laid open in a flat, in a plaine Map'. The first, containing Europe, Asia and Africa, and once thought to constitute all the world, corresponds to 'the literall, the Historical sense of the words'; the second, 'that of *America*' to 'an emergent … an occasionall sense of them'.[54] The image betrays a perhaps conscious acknowledgement that the discovery of America licenses and necessitates reinterpretations of a Scripture once thought absolute and literal. Probably *un*conscious is the implicit admission that America and its inhabitants are as unwelcome, treacherous and subversive of Christian order and authority as the 'humors … and rumors of men' which prompted James's 'Directions'. America is not so much a hemisphere as a '*demi-monde*', whose illegitimate presence is tacitly feared and resented.

In considering a third feature of Donne's sermon I want, however, to argue that, rather than evading them, Donne in fact positively harnessed, in a transfigured and redirected form, the disquieting but intoxicating energies of the New World. It seems evident, both from the sermon's direct confrontation of America as a theological conundrum, and from certain of Donne's other statements on religious belief, that he was temperamentally disinclined to settle for a faith which could not rigorously incorporate all natural phenomena, however novel and incongruous.[55] It might be said that, in 1622, just as in his colonial poems, Donne rose to the challenge of America's otherness and liminality because he was equipped with a correspondingly forceful psychological energy of his own. Particularly in the case of the two 'colonial elegies' – 'Loves Progress' and 'To his Mistress Going to Bed' – the vigorous, exploratory masculine aggression and self-confidence of the young poet more or less matches – and indeed is probably inflated by – the conspicuously maverick, guerilla-style colonialism of the 1580s and 1590s. By the time of the Virginia sermon, however, Donne was 50, and the colonial enterprise a wearier and less glamorous one. In Virginia especially the relatively transient hardships of the mariner or adventurer had been exchanged for the no less perilous, but more committed, endurance of the colonist and farmer.[56] The strangeness of America seemed hardly less diminished, and could scarcely have been mitigated by the sudden inexplicable hostility of its inhabitants. How, then, did Donne familiarise and control its potent energies at this stage?

The crux of Donne's Virginia Company address involves two related points. One is that the fulfilment of Christian eschatology requires the gospel to be '*preached to all the world*'; the second, that this had not been accomplished by the Apostles, and that, consequently, the colonial ministers of Protestant England now go to perform 'an *Apostolicall* function'.[57] At this stage Donne re-emphasises the theological significance of America with a particular urgency and excitement:

> Before the ende of the worlde come, before this mortalitie shall put on immortalitie, before the Creature shall be delivered of the bondage of corruption under which it groanes … before al things shal be subdued to Christ, his kingdome perfited, and the Last Enemy Death destroied, the Gospell must be preached to those men to whom ye send; to all men.[58]

Not only does Donne discover, in America, a new field in which to exercise Protestant virtues of industry, fortitude and zeal, but he in fact offers a notably Protestant opportunity to actively precipitate the grand finale of Christian history: 'Further and *hasten* you this blessed, this joyfull, this glorious consummation of all, and happie reunion of all bodies to their Soules, by preaching the *Gospell* to those men.'[59] By doing so, they will have 'made this Iland, which is but as the *Suburbs* of the old world, a Bridge, a Gallery to the new; to joyne all to that world that shall never grow old, the Kingdome of heaven' and will 'adde names … to the Booke of Life'.[60] Inscribing America within the 'Booke of Life', Donne might at first appear to have simply 'written it into submission' like so many of his contemporaries, subtracting from the world and adding to heaven. He has, however, also done something far more remarkable. By concluding the apostolic labour to be only partially complete, he reinserts the New World and its colonists within a newly-galvanised and potent tradition of Christian evangelism, almost seamlessly intergrafting biblical and contemporary history. The discovery of America, potentially an immense and irreparable breach in Christian cosmology, instead opens a *positive* space within the eschatological narrative, in which Jacobean Protestants, particular 'instruments of [God's] glory', can actively construct the last stages of the eternal kingdom.[61]

Using America as a bridge to future glory, rather than a rift driven through and against the grain of biblical history, Donne necessarily acknowledges the very strangeness and extremity of the New World. Such a vast, stubbornly alien presence needs to be refigured as a correspondingly urgent and drastic new phase in the drama of Christian eschatology. Although Donne, taking a carefully non-literal approach, explicitly rejects 'the *Carnall Kingdome* of the ... *Millenarians*', the sensuous force of his closing appeal ('Enamore them with your *Justice*, and ... inflame them with your *godlinesse*') carries an echo of the more extravagant, apocalyptic mood of William Symonds.[62] Symonds' 1609 Virginia sermon, owned by Donne (and perhaps attended by him), rises to the epistemological challenge of America by aligning it with the decisive events of Christian history. Such pivotal moments as the Flood, the burning of Sodom, the conversion of Constantine, and 'the desolation, and nakednesse of Antichrist, now readie to be cast into the fire' were, however, 'done in a corner, in comparison of that which is in hand'.[63] With militant fervour, Symonds proceeds to insist on the eschatological urgency of evangelical colonialism:

> Long since the Gospell of Christ did ride forth conquering that hee might overcome. And NOW, the hostes that are in heaven doe follow him on white horses. Now the Lord hath made bare his holy arme, in the sight of all the Gentiles; and all the ends of the earth shall see the salvation of our God ...[64]

The colonial enthusiasm of Symonds, who in 1605 had published the explicitly apocalyptic *Pisgah Evangelica*, cannot be unproblematically likened to that of Donne. The latter is very clear that the Kingdom of Glory cannot be understood temporally, and generally avoids the extravagance of tone found in more extreme evocations of the 'latter days'.[65]

Nevertheless, Symonds' own definition of 'the Millenaries' is those 'who looke for the gospell to be spread over all the world', and Donne is no less emphatic about the scriptural urgency of this task.[66] He appears, indeed, to perceive in America a kind of material embodiment of the fulfilment of Christian destiny. In a sermon attributed to the spring of 1623, he figures the unification of East and West by the pasting of 'a flat Map' upon 'a round body'. He then likens this fusion of opposites to 'a flat soule' which, applying its 'trouble to the body of the Merits, to the body of the Gospel of Jesus Christ' can thereby make 'thy West ... East, thy Trouble of Spirit ... Tranquility of Spirit'.[67] This image, in which the penitent sinner is implicitly asked to wrap their spirit around the body of the Saviour, seems to employ both Christ's material body, and the terrestrial globe, as attractively solid, definite and three-dimensional foci for devotional energies. In this case Donne does not specify America. About two and a half years later, however, he analogises the 'two Hemisphears, two half worlds' of a map with two key aspects of heaven: 'Halfe will be Joy, and halfe will be Glory; for in these two, the joy of heaven, and the glory of heaven, is all heaven often represented unto us.' Of these two hemispheres, 'the first hath been knowne long before, but the other (that of America, which is the richer in treasure) God reserved for later Discoveries'. Though the Almighty 'reserve that Hemisphear of heaven, which is the Glory thereof, to the Resurrection, yet the other Hemisphear, the Joy of heaven, God opens to our Discovery … in this world'.[68] Here Donne, who in 1622 expressly links universal evangelism with 'this *glorious* consummation of all', again associates the 'unripe', un-in-formed America with the future 'Glory' of 'the Resurrection'.[69]

Although in the later reference he attempts to shift the New World further into the spiritual plane, such attempts often only thinly mask the very stubborn, complex and massive physical reality of America. The glorious new hemisphere of 1625 had been, as recently as September 1622, an illegitimate, threatening and contingent new space, and Virginia clearly remained a highly topical, imaginatively vibrant presence in Donne's mind in the months before and after his sermon.[70] In joining the metaphorical 'East and West' he relies, notably, on 'the body of … Christ' – a conflation of spirit and matter which is not merely incorporated, but *personally* incorporated. Similarly, offering America to English Protestants as essentially 'the same *Stage*' as that once trodden by the Apostles, he attempts to join the alien new West with the Eastern, Old World, origins of Christianity by appealing to the active, pioneering spiritual individualism of the colonists. In this drama they are at once reassuringly authored and overseen by the Almighty, and yet implicitly granted the same degree of free will and personal expression available to the more assertive performers of the Blackfriars or Globe theatres.

While the delusion of 'assigning a certain time' to God's 'kingdome of glory' is effectively sidestepped, with America standing as a potent *spatial* expression of the coming 'last days', the actual time and places of late-Jacobean London and Virginia, and the absolute eschatological clock, nonetheless blur and strike together with conspicuous urgency.[71] In stepping into America, England's missionaries are unconsciously seen as stepping into an appealingly materialised, hopeful and liberating Christian future. This spatialised version of Christian history is as vast, dramatic, mysterious, fearful and – crucially – real as the last acts of the Scriptural narrative; or, indeed, as simultaneously real and unreal as Ralegh's tantalising quest for El Dorado.[72] Donne's religious celebration of America's liminal and alien expanses, if ostensibly very different from the sexualised and egotistical exhilaration of his poetry, is no less vigorous or triumphant.

For all this, Donne seems never to have finally subdued the anxieties raised by America. In his sermon of Christmas 1621, the 'naturall man' had been transformed from a potentially alien, unassimilable other into a supposedly neutral, experimental testing ground for the persuasiveness of Christian truth – a kind of theological litmus paper. Entering the field of New World evangelism on these apparently empirical terms, however, Donne necessarily opened up Christianity to correspondingly empirical subversions and dilemmas, in a realm whose physical realities repeatedly countered textual dogma.[73] It is surely not accidental that, in 1629, many decades after he had so rapidly and effortlessly thrown the Scholastic world-map around the globe of a teardrop, Donne effects a striking reversal of a pet image, referring now to 'That earth which is too much for man yet (for, as yet, a very great part of the earth is unpeopled)' and 'which, if we will cast it all but into a Mappe, costs many months labour to grave it'.[74] In theory Jacobean colonisation *should*, for English Protestants, have been an especially testing and arduous process. In practice, however, that power of dominating, possessing and psychologically fixing a shifting reality which had so often attracted Donne to cartography, had evidently given way to awe at the ungraspable strangeness and scope of a New World both resistant on its own terms, and fiercely contested by the powers of Spain and Portugal.[75] Five years after the dissolution of the Virginia Company, and with his own son, George, exposed to the perils of Spanish fleets in the Caribbean, Donne perhaps felt that America had added to the world in ways increasingly beyond his control.[76]

Notes

For advice and encouragement during the composition of this paper I am very grateful to Jerry Brotton, Ruth Gilbert, Claire Jowitt, Louise Leigh, Hugh McKay, Richard Moncrieff, Rebecca Rogers, Inga-Stina Ewbank, Carola Scott-Luckens and Andrew Skiller. For their long-term support and patience, I would like to thank Joan and Christopher Sugg, and Lesley and Douglas Fielding.

1. Donne, J. (1975), *The Complete English Poems*, ed. A.J. Smith, Harmondsworth: Penguin, p. 89.
2. See for example: Acosta, J. de ([1604] 1880), *The Natural and Moral History of the Indies*, trans. Edward Grimston, ed. C.R. Markham, 2 vols, London: Hakluyt Society, vol. 1, p. 40 (the original Spanish edition was published at Seville in 1590); and Donne, J. (1953–62), *The Complete Sermons of John Donne*, ed. G.R. Potter and E.M. Simpson, 10 vols, Berkeley and Los Angeles: California University Press, vol. 4, p. 279.
3. Aristotle (1984), 'Politics', trans. A. Platt, in *The Complete Works of Aristotle*, ed. J. Barnes, 2 vols, Guildford: Princeton University Press, vol. 1, pp. 21–3.
4. Hanke, L. (1959), *Aristotle and the American Indian: A Study in Race Prejudice in the Modern World*, London: Hollis & Carter, pp. 38–73.

5. Las Casas, B. (1583), *The Spanish Colonie, or Brief Chronicle of the Acts ... of the Spaniards in the West Indies*, trans. M.M.S. London; *Narratio Regionum Indicorum* (1614), Oppenheim.
6. See Maltby, W.S. (1971), *The Black Legend in England: The Development of Anti-Spanish Sentiment, 1558 1660*, Durham, NC: Duke University Press, esp. pp. 2–28.
7. On soul and body, see Aristotle (1984), 'Politics', pp. 33–4, and pp. 6–10; on male/female, ibid., pp. 12–15.
8. Ibid., pp. 20–26.
9. Ibid., p. 34; Aristotle (1984), 'De Anima', p. 5 and pp. 2–3.
10. Zamora, M. (1990–91) 'Abreast of Columbus: Gender and Discovery', *Cultural Critique*, **17**, 127–49.
11. See Donne (1953–62), vol. 4, p. 332, Whitehall, 1st Friday in Lent, 1622/3: '... the soul is the form of man'; and pp. 350–51; and vol. 9, p. 357, n.d.
12. Donne, J. (1980), *John Donne: Paradoxes and Problems*, ed. H. Peters, Oxford: Clarendon Press, pp. 28–9. Verse letter to the Countess of Huntingdon, ll. 1–2 (see Genesis 2:7); Milgate attributes this to the years 1608–09 (Donne, J., 1967, *Satires, Epigrams and Verse Letters*, ed. W. Milgate, Oxford: Clarendon Press, p. 247).
13. Donne (1953–62), vol. 4, p. 241, 'Anniversary Celebration of our Deliverance from the Powder Treason', 5 November 1622.
14. Verse Letter 'To the Countess of Huntingdon', Donne (1975), p. 236, ll. 9–10.
15. Donne (1953–62), vol. 9, p. 176, to the King at Whitehall, 12 February 1629/30, italic mine.
16. Bald, R.C. (1970), *John Donne: A Life*, Oxford: Clarendon Press, p. 162.
17. See Johnson, S. (1947), 'John Donne and the Virginia Company' *English Literary History*, **14**, 127–38, and Simpson, E.M. (1924), *A Study of the Prose Works of John Donne*, Oxford: Clarendon Press, p. 317. I am much indebted to the scholarship of Johnson's article for information on both Donne and Christopher Brooke.
18. Symonds, W. (1609), *Virginia. A Sermon Preached at White-Chappel ... 25 April 1609*, London, 8 May 1609. Keynes, G. (1973), *A Bibliography of Dr. John Donne*, Oxford: Clarendon Press, 'Books From Donne's Library', L177. On Company acquaintances, see Johnson (1947), p. 128. See also Gosse, E. (1899), *The Life and Letters of John Donne*, 2 vols, London: William Heinemann, vol. 1, p. 240, To Sir H. Goodyer, n.d.
19. 'That unripe side of earth', Donne (1975), pp. 238–41. There is no certain date for this letter; Milgate suggests 'soon after' 1605 (Donne, 1967, pp. 242–3).
20. 'To have written then', Donne (1975), pp. 227–9, 67–8; both Smith (Donne, 1975, p. 549) and Milgate (Donne, 1967, p. 262) attribute this to 1609.
21. It was in fact precisely on the issue of new stars that Donne himself explicitly questioned Aristotle (Donne, J. (1930), *Biathanatos*, New York: Facsimile Text Society, p. 146).
22. Kingsbury, S.M. (ed.) (1906–35), *The Records of the Virginia Company of London*, 4 vols, Washington, vol. 2, pp. 20, 26, 76 and 88–9.
23. Donne is listed as present on: 4 February 1622/3 (at morning and afternoon sessions); 5 February 1622/3; 5 March 1622/3; 7 May 1623; and 14 May 1623. See Kingsbury (1906–35), vol. 2, pp. 225, 231, 244, 300, 391 and 422.
24. See Donne (1953–62), vol. 3, pp. 357–9; vol. 4, pp. 59, 110, 181 and 350–51; vol. 6, p. 59.

25. Ibid., vol. 3, p. 357.
26. See ibid., vol. 4, p. 281; and, on James I's renewal of interest in the evidently rejuvenated Company of 1619, Kingsbury (1906–35), vol. 1, p. 24.
27. Waterhouse, E. (1622), *A Declaration of the State of the Colony in Virginia ... with a Relation of the Barbarous Massacre ... 22 March last*, London, p. 43, 21 August 1622.
28. Johnson (1947), pp. 130–31.
29. See Waterhouse (1622), pp. 15 and 24; Johnson (1947), pp. 131–3. I have quoted from the extracts of Brooke's very rare poem (11 September 1622) cited in Johnson (1947), p. 133.
30. Donne (1953–62), vol. 9, p. 176; Brooke cited in Johnson (1947), p. 133.
31. Donne (1953–62), vol. 4, p. 274.
32. Lim, W.S.H. (1998), *The Arts of Empire: The Poetics of Colonialism From Ralegh to Milton*, London: Associated University Press, pp. 79–80.
33. Donne (1953–62), vol. 4, p. 271; Brooke cited in Johnson (1947), p. 133.
34. Ibid., vol. 3, pp. 357–8, St Paul's, Christmas Day, 1621.
35. Ibid., vol. 3, pp. 358–9.
36. Ibid., vol. 3, p. 358.
37. Lim (1998), p. 78.
38. Donne (1953–62), vol. 4, pp. 269, 273.
39. Ibid., pp. 274–5.
40. Ibid., p. 269.
41. Ibid., p. 271.
42. Ibid., p. 279.
43. Ibid., p. 279: Matthew 24:14, Luke 24:47, Mark 16:20, Colossians 1:5, Romans 1:8, 16:19.
44. Donne (1953–62), vol. 4, p. 279.
45. Ibid., p. 279: Luke 2:1.
46. Donne (1953–62), vol. 4, pp. 279–80.
47. See Jowitt, C. (1995), 'Radical Identities? Native Americans, Jews and the English Commonwealth', *The Seventeenth Century*, **10**, 101–19.
48. Strachey, W. (1849), *The Historie of Travaile into Virginia*, ed. R.H. Major, London: Hakluyt Society, pp. 44–7; Acosta (1880), vol. 1, pp. 67–9, pp. 60–61. On the probability that the Indians 'descended from the cursed race of Cham' see also Strachey (1849), pp. 46–7, and Alderman Johnson in Kingsbury (1906–35), vol. 2, p. 397.
49. See Strachey (1849), pp. 44–7.
50. Donne (1953–62), vol. 4, p. 279.
51. Acosta (1880), **1**, p. 59.
52. Donne (1953–62), vol. 4, p. 279; vol. 9, p. 173, Whitehall, n.d.. Cf. Donne, J. (1952), *John Donne: Essayes in Divinity*, ed. E.M. Simpson, Oxford: Clarendon Press, p. 84: 'Jo. Acosta de procur. Ind. Sal. l.2 c.9'; and Donne, J. (1969), *Ignatius His Conclave*, ed. T.S. Healy, Oxford: Clarendon Press, p. 87.
53. Donne (1953–62), vol. 4, p. 59; Acosta (1880), vol. 1, pp. 19–23.
54. Donne (1953–62), vol. 4, p. 181, 15 September 1622, published November 1622.
55. See, for example, Donne (1953–62), vol. 3, p. 357, Christmas Day 1621: 'It is but a slacke opinion, it is not *Beliefe*, that is not grounded upon reason'; and vol. 4, p. 351, ll. 193–237, Easter day, 1623. Both these examples are in fact directly linked to the question of New World evangelism.
56. For the increased emphasis on settlement after 1619, see Kingsbury (1906–35), vol. 1, p. 39.

57. Donne (1953–62), vol. 4, pp. 279–280.
58. Ibid., p. 280.
59. Ibid., p. 280, first italic mine.
60. Ibid., pp. 280–81.
61. Ibid., p. 282.
62. Ibid., pp. 270, 280.
63. Symonds (1609), A2v.
64. Ibid., A2v–A3r.
65. Donne (1953–62), vol. 4, pp. 269–70.
66. Symonds (1609), p. 47.
67. Donne (1953–62), vol. 6, p. 59, pp. 1–2.
68. Ibid., vol. 7, p. 69, 29 January 1625/6.
69. Ibid., vol. 4, p. 280, italics mine.
70. Ibid., vol. 4, p. 181.
71. Ibid., vol. 4, p. 270.
72. See Ralegh, W. (1596), *The Discoverie of the Large, Rich and Bewtiful Empyre of Guiana*, London, n.p.
73. See especially Acosta (1880), vol. 1, pp. 20, 90.
74. Donne (1953–62), vol. 9, p. 47, April 1629. See also Letter to Robert Ker (Gosse, 1899, vol. 2, p. 191); Potter and Simpson date this between March 1622/3 and July 1623 (Donne, 1953–62, vol. 6, pp. 1–2).
75. On Donne and maps, see Lim (1998), pp. 65–7.
76. George Donne was in fact taken hostage by the Spanish in autumn 1629, escaping only after his father's death (Bald, 1970, pp. 552–3).

CHAPTER FOURTEEN

Alternative Planet: Kepler's *Somnium* (1634) and the New World

Mary Baine Campbell

Introduction: fiction as trouble

> What are for us among the main features of the entire universe – the 12 celestial signs, solstices, equinoxes, tropical years, sidereal years, equator, colures, tropics, arctic circles, and celestial poles – are all restricted to the very tiny terrestrial globe, and exist only in the imagination of earth dwellers. Hence, if we transfer the imagination to another globe, we must conceive of everything as changed.[1]

To conceive of everything as changed: the avocation of religious mystics, innovative philosophers, revolutionaries, utopia writers, and seventeenth-century geographers and astronomers. Such a task – 'we must conceive of everything as changed' – itself bespeaks a change, on 'this very tiny terrestrial globe' that Kepler in the *Somnium* describes as the most beautiful object in the lunar sky. We all know that 'change' was uncomfortably pervasive in early-modern Europe, though part of Kepler's charm is his very great enjoyment of it. Others enjoyed it less.[2] However alternative to itself the tiny terrestrial globe had become by the first decades of the seventeenth century, with its America and its Protestant churches, its venture capitalism and its vernacular literature, narratives of a possible alternative world seem to have been under a curious kind of ban. This was partly internal to the authors of such works, and partly made visible externally in the treatment of certain Italians who had not properly administered their own repression: Campanella all those long years in prison, Bruno dead at the stake, Galileo imprisoned and forced to recant. Kepler's little book created grave troubles even before its posthumous publication (1634) – for a long time it was famous mostly as the provocation (in pirated manuscript) of his mother's imprisonment and trial for witchcraft.[3]

Writers of fiction are in trouble again in our time, which bears some structural similarities to Kepler's – we too are in the midst of a revolution in what we now call 'information technology', and therefore in our notions of both 'the real' and the nature of representation, and we too seem to be enduring an 'age of religious wars' like the Thirty Years War that darkened and complicated Kepler's later career. The *fatwah* on Salman Rushdie is only the most notorious of many manifestations at the crossroads of these circumstances. But my interest in Kepler's strange little hybrid of science and fiction is not based in any historical homologies.[4] The *Somnium* appeared at a moment before the so-called 'scientific revolution' had drawn up its constitution and institutionalised its values. Though it had no real successors among scientific texts, it nonetheless points to a possibility, a road not travelled – down which we may see much of the only partly-mapped territory of modern science as a history and a kind of consciousness, and down which we may also catch a glimpse of what makes fiction so potentially powerful and subversive. It is worth wondering why this consciousness and this power were severed so early.

Troubles like Kepler's, and those that kept Bishop Godwin's *Man in the Moon* (1638) and Cyrano de Bergerac's *Voyage dans la Lune* (1657) out of print until their authors were dead (and even then the latter was abridged and bowdlerised), have been attributed to resistance to Copernicanism and the theological difficulty posed by several human races – for example, did Adam fall on each inhabited planet? Was he redeemed each time?[5] But there is more to the story. After all, Kepler did publish his *Conversation with the Starry Messenger* (1610), an enthusiastic response to Galileo's telescopic discovery of the moons of Jupiter and the mountains of our own moon – in itself a serious challenge to the classical dogma in which heavenly bodies were perfect and smooth. And, in 1543, Copernicus had published the *De revolutionibus* itself, dedicated to the Pope, which circulated unprohibited for almost 75 years. This is not to dismiss the sufferings of Galileo, but only to suggest that institutional denial of Copernicanism was not *total* and did not lead *invariably* to censorship. In 1638, the Anglican bishop John Wilkins published a Copernican account of the moon,[6] similarly emphasising its potential for supporting intelligent life, without a hint of reluctance or trouble. *A discovery of a world in the moon* was a prose discourse based on 13 propositions supported by logic, authorities and some credible data. The books their authors kept out of circulation, on the other hand, were structured as fictions.

Modern commentators on seventeenth-century moon voyages, and especially on Kepler's, usually point to their fictional frameworks as 'sugar-coatings' for the pill of controversial and difficult new ideas. But, as Edward Rosen makes clear in the Introduction to his English translation, the *Somnium*'s intended audience was astronomers, if not necessarily professional ones. Why would astronomers need their lunar geography sugar-coated? And, if fiction sugar-coats, why was it their *fictional* works that had to be posthumous? (All of Kepler's works were Copernican, not to mention difficult, but only *The Dream* was posthumously published.) I think it is disrespectful of fiction, even fiction at its most skeletal, to think of it as a palliative add-on or pedagogical strategy. It certainly leaves nothing to talk about, and indeed surprisingly few people have had anything interesting to say about the *Somnium*.[7]

Kepler's *Dream* and its various objects and means occupy the position of alternative to a large number of approved perspectives, genres and phenomena including, as we will see, its own. That some of the alternatives it represents were anticipated as dangerous by reactionary forces in its time suggests it is something more than a scientific or literary sport, the queer hot potato that has been tossed so unproductively from discipline to discipline since its first strong misreading put Frau Kepler in danger of her life. That to some degree it shared its troubles, not only with Copernican science but with other fiction, and particularly with other alternative worlds, supports this suggestion of cultural importance.

The context of New World voyages

Importance, we had better ask, of what?

The main text of the *Somnium*, the narrative itself (as opposed to the Notes – comprising 223 notes, plus Appendix and 38 notes on the Appendix – with which the original narrative was published) situates itself quickly at the intersection of two familiar kinds of narrative: the dream vision, so often launched, as this one is, by the narrator's falling asleep over a book; and what might seem almost its polar opposite, the genre known in those early days of exploration and empire-building as the 'voyage'.[8] 'It happened one night that after watching the stars and the moon, I went to bed and fell into a very deep sleep':[9] the dream is the quintessential form of *fiction*, making none of the historiographical claims of epic or even of romance, announcing itself as nothing if not other than what it tells you or says.[10]

> 50,000 German miles up in the ether lies the island of Levania. The road to it from here or from it to this earth is seldom open. When it is open it is easy for our kind (i.e. daemons), but for transporting men it is assuredly most difficult and fraught with danger.[11]

The whole *frisson* of the voyage is that it 'artlessly' details an actuality not just stranger than fiction but triumphantly alien to it: of course, Kepler did not stand first in this crossroads – even Dante did not – but his *Lunar Astronomy* is certainly a limit case for fiction, and it is highly, if inconsequentially, original for modern science. For that generic self-estrangement, that *alter et idem* structure, is exploited in every aspect and dimension of this book: it is, one might say, the take-home message. When did we last read a scientific text with a *message*?

Over and over, in fact, Kepler in his notes insists on this point – that he has a message. The Notes, apparently an exegetical alternative to the 'allegory' (as they label the main narrative), fuse another polarity: the theorists of the seventeenth century had advocated an almost hysterical purge of figure and polysemy from the language of scientific discourse, so the conflation here of allegory and high geometry is faintly eerie, despite the Neoplatonic vogue of numerology.[12] The conflation as such is highlighted by the presence in both main text and Notes of both allegorical narrative and geometric astronomy. At any rate the Notes say plainly on several occasions: 'Here you have the principal thesis', and that thesis is reiteratedly some form of the statement in Note 146: 'Everybody screams that the motion of the heavenly bodies around the earth and the motionlessness of the earth are manifest ... *To the eyes of the lunarians, I reply*, it is manifest that our earth, their Volva, rotates but their moon is motionless'.[13]

Some voyage narratives and histories of the New World, or works responding directly to them, had made this relativist point before (as of course had theorists on perspective in drawing): Las Casas loved the reversal of terms (his Spaniards in America were cannibals and 'savages'), as did a (temporarily) embittered Cabeza de Vaca, and Montaigne need hardly be mentioned again in this context.[14] Such inversion already had a long history in orientalist writing, and a popular adumbration in certain ritual enactments of 'the world upside down', such as the Feast of Fools, when children, animals and the minor clergy took over the cathedral to celebrate a Mass.[15] Somehow, though, the point runs wild this time, jumping the fences generally put round it of literariness or, in a more modern sense than Kepler's, of allegory. The book frightens people; it causes trouble of unintended sorts.

In February 1600, Giordano Bruno was burned at the stake, in part, some scholars propose, for teaching that there could be many worlds. (A world, of course, is something different from a planet – everyone knew there were several 'planets' – a world is *alive*.)[16] It is not much explanation to say that his idea 'contradicted' 'the Bible': so did many structural features of early-modern European cultural life – monogamy, for instance – and the Bible contradicts *itself* for those who are able to read. Something made it much harder for Bruno, and the slightly more cautious Kepler, than it had been in the fifteenth century for Nicholas of Cusa to present the possibility of worlds beyond the earth (or, in the early sixteenth, Marcellus Palingenius Stellatus, for whom the inhabitants of the other worlds were perfect and not in need of an absurdly repetitive Redeemer).[17] My thought is that the context for Bruno's execution was in part the new salience of alternative worlds *on* earth. Theologically the 'New World', with its populations for so long unevangelised by Christ's gospel, presented the same challenge as Bruno's many worlds – and, like them, goes unmentioned in the Bible. But, after an initial flurry of academic dismissiveness, it could not be denied or ignored; it had, therefore, to be contained or, perhaps more accurately, merged, through colonial, missionary and mercantile appropriation, with the one world Christ had saved and the Pope now controlled.[18]

Kepler reproduces the new cosmographical context everywhere in both the narrative of the *Somnium* and its voluminous Notes. The latter are full of specific allusion to voyage literature, data from which are properly cited as if it belonged to the same technical literature to which Kepler's text putatively belongs. A continuous textual plane is thus constructed on which Peru and Levania (as he calls his lunar world) are 'islands', both being shielded from excessive sun by low cloud cover. Lunar cold is colder than the American 'Quivera' (José de Acosta's Kansas, linguistically distorted and geographically misplaced).[19] The 'island' is the land form that functioned as a kind of mastertrope of New World topography, and that characterised the focus of classic voyage literature, especially where it spoke most directly to private desire: Columbus finds islands, and Thomas More, and Andre Thevet. *The Purple Island, The Isle of Pines*, the islands of Marguerite de Navarre and of Shakespeare, Defoe and Swift: we can follow in these dots a trajectory that points to richly imaginary fulfilments. In New World writing this meeting or mutual generation of fulfilment and desire can be diverted to the socially and politically useful. 'Adventure capitalism' after all requires reachable countries of the heart's desire.[20]

In fact, Kepler did think space travel was plausible, that the moon was reachable, and assumed that Germans would get there first. Bishop Wilkins was of the opinion that the moon would lose her political virginity to a British flag. Neither of them thought it would take 350 more years of research and investment, and Wilkins (cousin to Oliver Cromwell) even got his government interested.[21] But it wasn't good business, an island '50,000 German miles up in the ether'; and this meant no containment or practical diversion of the desires provoked by our readerly contact with the planet of madness. Where Columbus or Thevet could conclude or even foreclose a description of New World topography by remarking that an island or bay would be 'easily fortified' to ensure control of nearby resources, Kepler punctuates his narrative with variations on the less profitable refrain of changing places in the mind: 'For Levania seems to its inhabitants to remain just as motionless among the moving stars as does our earth to us humans'.[22] The business of narrative is perception, not acquisition: a mobilised, alterable, positional kind of perception rather than the steely focus of the 'imperial eye'.[23]

Moon voyages, then, present (or *can* present) alternative worlds that offer most saliently the radical fact of Alternativity itself. And they do this vis-à-vis an 'island' which, unlike America or even Iceland (the frame setting of Kepler's *Somnium*), is perfectly visible almost every night, to everyone. Kepler's voyage, ostensibly the least satirical or political of the lot, got him in the most trouble of any of the moon writers, and more than once. These troubles generated the enormous expansion represented by the Notes, begun about ten years after the original text and amounting to roughly six times the length of the original narrative. The Notes augment the scientific data already foregrounded in the main body of the narrative and defend various ludic moments against their ludicrous misreading in the events of his mother's imprisonment and trial. They function as an exegesis of the narrative Kepler repeatedly terms an 'allegory', suggesting that scientific discourse might have been seen as a kind of mediation, even arbitration, between suggestive new empirical data and the mainstream cosmology of the educated – and not so educated – public. (This notion of arbitration is fostered by Kepler's frame narrative, with its weird opening reminder of the civil war brewing in Bohemia between the Holy Roman Emperor Rudolph and his ambitious and angry brother[24]). The rest of this essay will add another layer of exegesis to Kepler's own: we have become used to lunar astronomy and the solar system it depends on, but we have forgotten the kind of science that offers its discourse as allegorical and its object as poetic. What does this internally alternative text, at once fiction and science, narrative and interpretation, have to tell the modern exegete about the meaning of the moon and the nature of early-modern science?

The art of Kepler's science

The chief values of this underground 'Astronomy's' lunar world are alternativity and inclusivity. The title, *Dream, or Lunar Astronomy*, means *both* Dream *and* Lunar Astronomy, despite the generic opposition of these forms. To quote another seventeenth-century title, that of Joseph Hall's contemporary English imaginary voyage, this lunar world is *alter ET idem* (other and yet the same) – not in the predictable and usually reactionary mode of satire but in a way that could have changed both dream literature (even the very theory of dream signification) and science.[25] It is almost impossible at this late point in the history of modern science to imagine a technology of thought that might unscandalously include either the data or signifying structures of so ontologically opposed a form of understanding as the dream. But early-modern natural philosophers such as Cardano, Descartes and Henry More attest in published work to dreams in which they solved major difficulties of practice and theory. And the dream vision was a genre closely linked to pronouncements of divine authority.

The strange first paragraph of Kepler's narrative opens up two more pairs of alternatives: who would expect a book with either of the titles quoted above to begin like this: 'In the year 1608 there was a heated quarrel between the Emperor Rudolph and his brother, the Archduke Matthias'? The strangeness of this starting point is emphasised by the absence of any segue between the topic of Bohemian civil strife and the dream vision's pivotal formula; 'I went to bed and fell into a very deep sleep.' We have been offered the picture of a *mutually exclusive* alternativity between two nearly identical persons, in the royal brothers' struggle for a single throne; Kepler (the frame narrator) then starts reading about the strikingly parallel troubles of the legendary Bohemian Queen Libussa and falls asleep over his book. His dream seems to have been provoked by thoughts of the mutual exclusivity of contenders for power in the political realm. It offers the Moon as an alternative to Bohemia – the night-time world of dream, moon, the perennial and, as we will see, the uncontentious, replaces the daytime political world of strife and change; the book Kepler *dreams* he is reading about the young astronomer Durocotus and the Moon transforms the book about Libussa and Bohemia that he had 'really' been reading in the frame narrative.

The one protagonist does not simply replace the other, however. Libussa – sorceress, beloved ruler and dynastic mother of Bohemia – seems less an alter ego of our author narrator (or of his dream self, Durocotus) than of Durocotus' Icelandic mother, Fiolxhilde: both figures are sorceresses, mothers, female authorities. Why does this dream mother die as soon as Durocotus (narrator of the book-within-a-book that Kepler reads in his dream) introduces her? 'Her recent death freed me to write'[26] is not a surprising statement at the level of psychological realism, nor in the context of Kepler's exegetical equation of Durocotus' mother with 'Ignorance'.[27] But Kepler's equation doesn't actually hold up very well: the mother, supposedly Ignorance, is described right away as tactically wise, learned in herb-lore, scornful of 'vicious people who malign what dull minds fail to understand',[28] and possessed of a key piece of data withheld from Durocotus – his own father's name. She seems to know quite a bit, but it is an alternative knowledge (soon to pass away as both contents and a relation to knowing, though Kepler may not have anticipated that).[29] In fact, almost the entire text (sans Notes), subtitled 'The Lunar Astronomy', is a transcript of oral, daemonic lore to which Durocotus' mother provides him (and therefore us) the access. When the son returns from his long apprenticeship with Tycho Brahe she tells him that now she can die, 'since she was leaving behind a son who would inherit her knowledge, the only thing she possessed'![30] A serious conflict is registered here where this figure, of a kind of knowledge still highly valuable *to a scientist* (Kepler, Durocotus, his astronomer readers), must die before the scientist is free to write. The mother's knowledge is the writing's content, at the same time as the mother is its censor.

Kepler's appetite for the hybrid, his ability to tolerate ambivalence, are high – higher perhaps than a fully rationalised science can express. Duracotus, Kepler's stand-in in the dream, has spent the years of his adolescence away from Iceland and his mother, studying with Imperial Mathematician Tycho Brahe (as had Kepler himself) on the then-Danish island of Hven. On this lush, temperate island he learns Danish quickly from Brahe's other students, with whom he exchanges oral geographies and from whom, as well as from Brahe, he learns the techniques of astronomy. Again, his mother is included in the network of astronomical knowledge: 'This practice reminded me of my mother, because she, too, used to commune with the moon constantly'.[31] When he returns from his earthly travels and adventures and tells his mother of them, she gives voice to one potential consequence of reading or hearing about alternative worlds: 'We [here in Iceland] are burdened with cold and darkness and other discomforts, which I feel only now, after I have learned from you about the salubriousness of the other lands.'[32] Such representations teach desire and dissatisfaction with one's 'natural' home. The mother is depicted in relation to a knowledge and a form of representation disconnected from those of the 'New Science'. The text's fictional mother's connection with the lore and practice of 'wise women' had fatal consequences for Kepler's real mother; thus oddly supporting the concept Fiolxhilde embodies of language – which summons spirits, constitutes travel and jails old women – as magically productive and performative rather than passively descriptive.[33] Although her son Durocotus has travelled in the body and she has not, she has her own teacher, her own kind of travel to offer, her own, or at least her teacher's, kind of representation – presumably alternatives to the travel and scientific education her son has just been narrating. Her teacher is a wise and gentle spirit,

> who is evoked by twenty-one characters. By his help [later the spirit is referred to as 'her'] I am not infrequently whisked in an instant to other shores … or if I am frightened away from some of them on account of their distance, by inquiring about them I gain as much as if I were there in person … I should like you to become my companion on a visit, particularly, to that region of which he has spoken to me so often.[34]

This teacher of Fiolxhilde's manages a powerful kind of representation: Durocotus' 'visit' to the moon, the main business of the text, will take the form of listening to him/her speak, in the ritual setting of a crossroads on the night of the new moon. This preternatural vivaciousness of mimesis brings to mind Michael Taussig's recent meditation on anthropological encounter, inspired by the mimetic practice of the Cuna Indians of Central America, during the period of their first contact with Western information technology.[35] His *Mimesis and Alterity* would easily absorb such a transportational notion of narrative as Fiolxhilde's; the book is obsessed with the way the magical practices of what he calls 'copy' (and we would call representation) and 'contact' (manipulation of actual substance of the object to be affected) slide into a single identity (the fingerprint is a good instance, as is the voodoo doll containing its object's nail clippings). Although Taussig is speaking mostly about the visual mimesis central to this non-alphabetic culture, he makes it clear that the Cuna do not have exclusive rights to this kind of thinking. They represent only one, strongly unanimous, example of the intensification of mimetic power felt in the encounter with new forms and objects of representation. The spiritual or 'pneumatic' magic of Ficino, Pico and Kepler's contemporary Bruno might be another.

Taussig, invoking Walter Benjamin's writing on the medium of film, spreads the transformational net of mimesis more widely when he speaks of 'the unstoppable merging of the object of perception with the body of the perceiver and not just with the mind's eye'.[36] Commentators have often complained of Kepler's sloppiness in labelling the narrator of the astronomical part of the *Somnium* the 'Daemon from [ex] Levania', since the daemon makes it plain that his native planet is not Levania but the Earth and that his discourse on the moon is that of a foreign traveller. But perhaps in this case, not only does the copy partake of the substance of the thing copied, but the copier too participates in the radioactive spread of mimesis: this astronomy is not only *about* the moon but, as the subtitle says, *lunar*, even a little lunatic. There is a strangely permeable membrane between the scientist and his object, here acquiring many of the characteristics of a subject.[37]

At several points Taussig connects the manipulations of Cuna magical 'copying' directly with the technical instruments and practices of modern science:

> By holding still the frame where previously the eye was disposed to skid, by focusing down into, by enlargement, by slowing down the motion of reality, scientific knowledge is obtained through mimetic reproduction in many ways. We see and comprehend hidden details of familiar objects. We become aware of patterns and necessities that had hitherto invisibly ruled our lives.[38]

The connection Taussig is making between the apparent atavism of sympathetic magic and the technological-scientific manipulation of perception is relevant here: the 'twenty-one characters' by which Fiolxhilde's demonic teacher is 'evoked' in the passage quoted above are the 21 letters in the title of Kepler's book, the *Astronomia Copernicana* (1618), according to one of his own notes. A peculiar economy: while the narrator has been away studying in Imperial Mathematician Tycho Brahe's observatory, his mother the 'wise woman' has also been 'away', transported by the *Astronomia Copernicana*. When the young astronomer comes home from his years studying the heavens with the Imperial Mathematician, it is his mother the witch who teaches him everything he has to tell us about the moon, through the intermediary of a spirit who seems in turn to be the animated text of a 'properly' scientific book by the narrator's alter ego and author, Johannes Kepler, *current* Imperial Mathematician and one time student of Brahe's![39] The relations between scientist-author, scientist-protagonist, protagonist's magico-scientific mother and the author's previous astronomical text are collapsed here into something like identity, one substance.

About half the text of the *Somnium* is constituted by lavishly detailed accounts of the geography and climatology of 'Levania', along with the disposition of the heavenly bodies as perceived from several places on the moon's surface. Technical as these portions may be, their presentation is grounded in the magnetising principles of desire and lack. The features of the moon and its skies are always represented as things *seen* by 'Subvolvans' or 'Privolvans' (the respective inhabitants of the two lunar hemispheres) – which lends an eerie emptiness to the landscape so described, as we are not introduced to these inhabitants until the last couple of pages.[40] This is, of course, an increasingly familiar arrangement of information in earthly voyage literature written in the context of colonial acquisition. As the discipline of anthropology emerged to testify, we do need to know about the Natives; but our desire to settle their land is more easily aroused if they aren't looking on as we read about it. In Kepler's narrative, on the other hand, their absence and the landscape's human emptiness irradiate the lengthy discussions of what is visible on their horizons; we are identifying vividly with lunarians, or at any rate seeing through their eyes, well before we know what they look like. And we are identifying, for the purposes of 'lunar astronomy', with *both* Subvolvans *and* Privolvans – groups which the frame of the *Somnium* suggests would be at perpetual odds on earth (like the Emperor and his brother, or Libussa and her barons).

The two hemispheres of earth are discussed as well, in a moment of breathtaking aestheticisation. Earth itself is alternative in this book – as 'Volva', the giant moon that stands in for the Moon on the Moon where the inhabitants are, as the Daemon puts it, 'completely deprived' of the Moon. The political geography (or 'selonography') of the Moon seems based on Volva as an object of not only aesthetic but also erotic adoration. It is decorated with sexual images, likened to a jewel, overwhelmingly specular – it is characterised above all by the definitive absence and difficulty or impossibility of access which marks off the erotic from the merely admired.

> [T]hey enjoy [it] to make up for our moon, of which they … are completely deprived. From the perennial presence of this Volva this region is termed the Sub-volvan, just as from the absence of Volva the other region is called the Privolvan, because they are *deprived* of the sight of Volva.[41]

Since Volva rotates daily on its axis, the Subvolvans tell time by the position of its 'spots' (continents): the Daemon chooses to describe the visible disk of the planet at a moment when familiar parts of both (in our case artificial) hemispheres are visible at once. The Old World ('the eastern side') 'looks like the front of a human head cut off at the shoulders and leaning forward to kiss a young girl in a long dress, who stretches her hand back to attract a leaping cat'. South America 'you might call ... the outline of a bell hanging from a rope and swinging westward'.[42]

These images are impressive acts of visualisation, as indeed is the whole astronomical narrative of the *Somnium*, coming from a man who, though he knew maps, did not grow up in a map-saturated culture like that of the twentieth or even the nineteenth century. There were no maps of the world on his classroom walls, his watch or his credit card, no poster-sized photographs in his dorm room of the Earth as seen from spaceships. He had not, by the time of the early drafts anyway, even seen a telescope. But what is crucial about the images is that they constitute the first detailed description of the planet Earth from an external vantage point: of Earth as a specular object. This is surely an occasion for vertigo, when, to quote Frederic Jameson (out of context) on the genre of romance, 'the *worldness* of *world* reveals or manifests itself, ... [or] *world* in the technical sense of the transcendental horizon of our experience becomes visible in an inner-worldly sense'.[43] What could it mean for such a moment, which feels to me even now like a terrifying loss of ground, to bear as its visual content the image of a young girl playing with a cat while a man tries to kiss her?

The image is meant to convey the *orbis terrarum*, with Africa the Othello leaning to kiss the innocently preoccupied Desdemona of Europe, the still unravished bride perpetually rushing westward, forever panting and forever – how does Kepler know this? – 'young'. On the other side of the world a bell is tolling mightily. This is an untidy reversal of the actual story, in which a predatory Europe delights in interrupting supposedly 'childlike' Americans and Africans at play for a rambunctious episode of rape and conquest. As Kepler points out in his Notes, he has in fact had to reverse the cardinal points in his description, because the lunar viewer of these moving 'spots' sees eastward movement as westward.[44] But whatever the allegorical narrative of these cameos, most salient and most significant is the fact of a world, 'the transcendental horizon of our experience', shrunk to a shiny medallion, an amusingly suggestive ornament hanging in the cabinet of some lunar Rudolph (Kepler's imperial employer was famous for his *Wunderkammern*).[45] Of all the imaginative place-changing in this text, the shift of our human world, staggering with the weight of its violent histories, now ineluctably visible and ornamental in the sky of another planet, seems the most revolutionary in its conception and presentation. Another such image, this time photographic, fed revolutionary thought and feeling in 1970: the translation of the sublime into the beautiful can be experienced as a great disburdening of the tragic sense of life, with its links to monumental and imperial history – although ironically that image became a sign of America's imperial grandeur as well as of the lightness and loveliness of this (un)sceptred isle, this Earth.

Lunar ecology

Science fiction as the 'extrapolative' art we know today probably begins with this sentence near the end of the *Somnium*'s main narrative: 'For in general the Subvolvan hemisphere is comparable to our cantons, towns and gardens, the Privolvan to our open country, forests and deserts'.[46] Why is this account of the non-existent lunarian ecosystem here? How does it fit in with an otherwise textbook accurate account of things visible and knowable? In every other feature of Kepler's selenography the thrill is that we're seeing up-close or from the Other perspective something plainly visible to us from childhood on. One answer is that the genre of the 'voyage' demands it – and as a parodist Kepler is more urgently required to obey than if he were writing an account of actual exploration. The lifeworld of a potential colony, however, was characteristically rendered in itemisable form, rendering the unusable aspects of it conceptually detachable. Kepler's brief account of lunar flora and fauna, augmented in the 'Selenographical Appendix' where he describes the Levanian method of building what we now call craters, makes size, shape, longevity, diet, habitat and climate mutually expressive and interdependent. This reads like unremarkable common-sense to modern students, but produced in the time of the Jesuit polymath Athanasius Kircher's attempt to recalculate the size and interior design of Noah's Ark so as to accomodate the fauna of the New World, it represents a less familiar and less easily commodifiable view of organic life.[47]

Kepler's lunar ecology blurs two other lines of identity: he does not specify human status for one or another lunar species, yet building activities alluded to in the narrative and dwelt on in the Appendix make it clear that some 'moon dwellers' are rational creatures (and thus of course redeemable, though he makes no mention of that).

> The Privolvans have no fixed abode … In the course of one of their days they roam in crowds over their whole sphere, each according to his own nature: some use their legs, which far surpass those of our camels; some resort to wings; and some follow the receding water in boats.[48]

And, despite the possession of reason, the arts and enough political organisation to build and live in fortified 'towns', those 'moon dwellers', capable of conquest and colonisation show no interest in it. Every month all the water on the planet is drawn to the Subvolvan hemisphere for two weeks by the combined attraction of the sun and Volva, but the hemispheres have not organised the conflict this would naturally provoke on earth. The fortification of the towns described in the Appendix protects them mainly 'from the mossy wetness' of this monthly deluge and 'from the heat of the sun'.[49] It is not just that the moon-dwellers don't think of military conflict or colonial aggression – Kepler doesn't mention these things himself. They are absent from his dream, as Volva is absent from our skies.

* * *

Kepler's *Somnium* is only a parody in the widest sense – its energies are devoted to other ends than the satirising of colonialist voyage literature or the subtle admonishment of kings. I have drawn attention to its performance of these latter functions as a way to highlight some of what a 'Lunar Astronomy' could include in the seventeenth century. Even now, almost four busy centuries after Kepler began to imagine and analyse the 'spots' on another world, in a time when science fiction is pervasive and human beings take photographs of each other on the moon, it is possible to be moved by Kepler's response to the 'worldness' of the moon. That it inspired him to write a book so fully fictional and scientific at once – that we are invited, as Taussig would say (thinking of the camera's 'enchanted glass') – to 'focus down into' the data of the new astronomy, is not the least of the inclusivities sponsored by his sense of the moon.

The abrupt ending of the narrative is part of the project of making science signify: as the Levanian Daemon is telling Duracotus about Subvolva's 'constant cloud cover and rain', 'a wind arose with the rattle of rain, disturbing [Kepler's] sleep and at the same time wiping out the end of the book'.[50] The sad way that deeply meaningful Dream Rain turns out to be nothing but arbitrary actual rain is a familiar let-down to anyone who has ever had a dream – the inevitable lack of closure in dream narrative always refers us to the incompatability between a world of meaning and a world of brute physics. The end of Kepler's *Somnium* replays that let-down, which 'wipes out … the end of the book': a stark reminder of what the loss of a signifying science might feel like. The underground possibility this book embodies involves an idea of science as a self-transformative practice of seeing, of *looking*. Not in order to decipher a future, as Kepler had been pressured to do in astrological work commissioned by Rudolph and others at court, but in order to 'change your life'. Or, as his countryman Kepler would say, to 'change everything'.

Notes

This essay began as a response to the Fourteenth Barnard Medieval and Renaissance Conference, 'Alternative Realities', which Christopher Baswell – impressario in many worlds – rightly thought I'd like to be part of. Since then it's passed through the sieves of the Science and Culture Seminar at Brandeis University (where physicist and historian Sam Schweber gave me much to think about) and an NEH-sponsored seminar connected to an interdisciplinary pilot curriculum at the University of Maine (Orono), where chemist and dancer François Amar was my inspiring host. James Romm hosted a visit to Bard College to talk to his undergraduate class about Kepler: the wonder and pleasure with which his students spoke of the text and its moon confirmed the sense of historical loss with which my essay concludes. I thank all these people, as well as Marie Plasse and Jason McLachlan, who helped me to say what I mean. Reprinted, modified and expanded from Mary Baine Campbell: 'On the Infinite Universe and the Innumerable Worlds', in *Wonder and Science: Imagining Worlds in Early Modern Science*. Copyright 1999 by Cornell University. Used by permission of the publisher, Cornell University Press.

1. Kepler, J. (1634), *Somnium, seu opus posthumous de astronomi lunari*, Zagan and Frankfurt-am-Main: n.p.
2. The *Somnium*, like so many seventeenth-century works depicting the moon and its imagined life, was published only posthumously. An earlier, shorter and less literary version of it submitted as a dissertation at Tübingen had been rejected by the professor in charge of arranging disputations. In a letter to a friend looking for a mathematics textbook, years later, Kepler jokingly suggested publishing his *Somnium* for the purpose: 'This will provide the fare for us who are being chased off the earth …' (quoted in E. Rosen (trans.) (1967), *Kepler's Somnium: The Dream, or Posthumous Work on Lunar Astronomy*, Madison and London: University of Wisconsin Press).
3. The full story is available in Max Caspar's biography, Caspar, M. (1993), *Kepler*, trans. C. Hellman (1959), revised by O. Gingerich and A. Segonds, New York: Dover. For more detail (and quite differently) see John Lear's monograph-length introduction to a modern English translation of the *Somnium* inspired by the space race of the 1960s: Lear, J. (ed.) (1965), *Kepler's Dream*, trans. P.F. Kirkwood, Berkeley and Los Angeles: University of California Press. In Lear's version the story is reminiscent of the troubles faced by women in Salem, Massachusetts, later in the century: a village feud found a use for Kepler's manuscript as evidence – amidst a wealth of the usual anecdotes about sick cows and sudden joint pains – of sorcery in the behaviour of garrulous and quarrelsome old Frau Kepler. The mother of the frame-narrative's protagonist is a herb-woman and folk astronomer, in contact with spirits who tell her about distant places they are able to visit, including the moon. Frau Kepler was in prison for two years during the proceedings, and although Kepler (who had to drop all other business in pursuit of her freedom) managed to win her release, she died shortly afterwards, worn down by the conditions of her ordeal.
4. Throughout the text of this essay I will be using the term 'science', despite its rather different reference in the late sixteenth and early seventeenth centuries, along with the less anachronistic 'natural philosophy' and so on: this is because I am writing in this case with an eye on the present as an outcome (or rather, not an outcome) of the early-modern matters here discussed.

5. Bergerac, C. de (1657), *L'histoire comique de les etats et empires de la lune*, Paris. Godwin, F.B. (1638), *The Man in the Moone*, trans. E.M., London. See, for example, Christianson, G.E. (1976), 'Kepler's *Somnium*: Science fiction and the renaissance scientist', *Science Fiction Studies* **3** (2), 79–90.
6. See Wilkins, J. (1638), *A Discovery of a World in the Moon*, London: n.p., p. 93.
7. Among the few exceptions, see the relevant chapters in Reiss, T. (1982), *The Discourse of Modernism*, Ithaca: Cornell University Press, and Hallyn, F. (1993), *The Poetic Structure of the World: Copernicus and Kepler*, trans. J. Hall, New York: Zone Books.
8. The dream vision is a literary genre of framed narrative very popular in the later Middle Ages and early Renaissance though known in many periods, sometimes as an account of an actual dream. Canonical late medieval literary dream visions include in the thirteenth century Dante's *Commedia* and Jean de Meun's *Roman de la Rose*, and in the fourteenth Chaucer's *Book of the Duchess*; the twentieth-century film of *The Wizard of Oz* is a more contemporary example. See Spearing, A.C. (1976), *Medieval-Dream Poetry*, Cambridge: Cambridge University Press; also Browne, P. (ed.) (1999), *Reading Dreams: The Interpretation of Dreams from Chaucer to Shakespeare*, Oxford: Oxford University Press, and Weidhorn, M. (1970), *Dreams in Seventeenth-Century Literature*, The Hague: Mouton. For the vision itself, see Le Goff, J. (1988), 'Part V: Dreams,' in *Medieval Imagination*, trans. A. Goldhammer, Chicago and London: University of Chicago Press, pp. 193–242.
9. Rosen (1967), p. 11.
10. I have consulted Frisch's edition of the *opera* for the Latin text: Frisch, M. (1857–71), *Joannis Kepleri astronomi opera omnia*, 8 vols, Frankfurt and Erlangen.
11. Rosen (1967), p. 15.
12. The periodical literature on seventeenth-century language theory, the search for a 'real character' and the purging of natural language for the purposes of natural philosophy is large. A good starting place would be Srigley, M. (1988), 'The lascivious metaphor: The evolution of the plain style in the seventeenth century', *Studia Neophilogica*, **60**, 179–92. For a more comprehensive introduction, see Slaughter, M. (1982), *Universal Languages and Scientific Taxonomy in the Seventeenth Century*, Cambridge: Cambridge University Press. Also relevant is Bono, J.J. (1995), *The Word of God and the Languages of Man: Interpreting Nature in Early Modern Science and Medicine*, Madison: University of Wisconsin Press.
13. Rosen (1967), p.106, my emphasis.
14. See especially Las Casas' moving propaganda work *Brevíssima Relación*, which is often credited with having begun the 'Black Legend' of Spanish cruelty in the New World: Las Casas, B. de (1552), *Brevíssima Relación de la Destrucción de las Indias*, Madrid. Also Cabeza de Vaca's *Relación* of his eight years' wandering, practising as faith healer among native peoples across the American Southwest and Mexico: Cabeza de Vaca, A. (1542), *La Relación*, Zamora; and Montaigne's 'Of Cannibals' and 'Of Coaches', Montaigne, M. de (1580 ff.), in *Essais*, Bordeaux.
15. See Davis, N.Z. (1975), 'The Reasons of Misrule', in N. Davis, *Society and Culture in Early Modern France*, Stanford: Stanford University Press; also Burke, P. (1978), 'The World Upside Down' in Peter Burke's *Popular Culture in Early Modern Europe*, New York: Harper and Row, pp. 185–90.

16. For more information on the troubling and troubled career of Bruno, see especially Yates, F. (1964), *Giordano Bruno and the Hermetic Tradition*, London: Routledge and Kegan Paul. I share Yates' sense of an underground culture in the Renaissance; the present essay suffers, perhaps, from a lack of attention to the hermetic Kepler. See also Singer, D.W. (1950), *Giordano Bruno: His Life and Thought, with Annotated Translation of His Work on the Infinite Universe and Worlds*, New York: Henry Schuman. This work contains Singer's own translation of *De l'infinito universo et mondi* (1584).
17. Palengenius is the author of *Zodiacus vitae* (*c*. 1531). On the long controversy over the theological and philosophical implications of the existence of other worlds, see Dick, S.J. (1982), *Plurality of Worlds: The Origins of the Extraterrestrial Life Debate from Democritus to Kant*, Cambridge: Cambridge University Press, especially chs 2 and 4.
18. On post-Reformation Catholic universalism and globalism, see Headley, J. (1998), 'Geography and empire in the late renaissance: Botero's assignment, western universalism, and the civilizing process'; and Headley, J. (1997), 'The sixteenth-century Venetian celebration of the earth's total habitability', *Journal of World History* **8** (1), 1–27. But the Reformation was not quite underway when Pope Alexander VI divided the globe between Spain and Portugal in the 1493 bull on which the Treaty of Tordesillas was based.
19. A significantly *exotic* comparison, especially given the setting of Kepler's daemon speaker and his interlocutor – out of doors on a dark night in Iceland! The mountains of native Bohemia could have served as well, but the moon is closer in emotional geography, at least for a seventeenth-century European, to Kansas/Quivera. For Quiveran cold, see Acosta, J. de (1596), *Historia natural y morale de las Indias*, Köln, bk 2, ch. 10 (available on microform in Goldsmith-Kress Library of Economic Literature, no. 274).
20. For 'adventure capitalism', see Michael Nerlich's Marxist account of literary change in a changing Europe: Nerlich, M. (1987), *Ideology of Adventure: Studies in Modern Consciousness, 1100–1750*, vol. 1, trans. R. Crowley, Minneapolis: University of Minnesota Press.
21. On early national space programmes, see Nicolson, M.H. (1966), 'The Discovery of Space', in O.B. Hardison (ed.), *Medieval and Renaissance Studies: Proceedings of the Southeast Institute for Renaissance Studies*, Chapel Hill, NC: University of North Carolina Press (pp. 54–5).
22. Rosen (1967), p. 17.
23. For an analysis of the structure and historical context of that focused gaze, see Pratt, M.L. (1993), *Imperial Eyes: Travel Writing and Transculturation*, London and New York: Routledge.
24. Rudolph II, the increasingly unstable king of Bohemia, and Holy Roman Emperor at the time of the first composition of the *Somnium*, was forced to abdicate in favour of his brother Matthias, Archduke of Austria, in 1611. Rudolph's Prague (then capital of the Empire) had been a haven for intellectuals and virtuosi, and Rudolph himself was a notable collector. Neither brother was able to fend off the impending conflict of the so-called Thirty Years War, which began in Prague in 1618, though both had negotiated with the Protestants and granted them religious freedom in their lands.
25. Hall, J. (1605), *Mundus alter et idem*, London. Alice Ingerson examines the dichotomy that has wedged apart Kepler's mental frameworks, from the point of view of a practising scientist and past editor of an interdisciplinary academic

journal (*Forest and Conservation History*): Ingerson, A.E. (1994), 'Tracking and testing the nature/culture dichotomy in practice', in C. Crumley (ed.), *Historical Ecology: Cultural Knowledge and Changing Landscapes*, Santa Fe: SAR Press, pp. 43–66.

26. Rosen (1967), p. 11.
27. See Kepler's Note 4: 'untutored experience or, to use medical terminology, empirical practice is the mother who gives birth to Science as her offspring. For him it is not safe, so long as his mother, Ignorance, survives among men, to reveal to the public the deeply hidden causes of things' (Rosen, 1967, p. 36).
28. Ibid., p. 12.
29. For the story of the changing of the guard, and of the content of people's knowledge, see Thomas, K. (1983), *Man and the Natural World: A History of the Modern Sensibility*, New York: Pantheon Books. Chapter 2 is especially relevant: 'Sir Joseph Banks, the future President of the Royal Society, as a schoolboy paid herb-women to teach him the names of flowers. Physicians and apothecaries had long depended for their supplies upon such persons, what William Turner called "the old wives that gather herbs"' (p. 73). The chapter goes on to detail the imposition of Latin names and taxonomies on the plants identified by herb-women, fowlers and ex-soldiers, so that eventually 'farmers who still used "vulgar, provincial names …" found themselves unable to communicate with the naturalists' (p. 87).
30. Rosen (1967), p. 13.
31. Ibid., p. 13.
32. Ibid., p. 14.
33. On relations between 'wise women' and early-modern natural philosophy, particularly in Kepler's case, see Reeves, E. (1999) 'Old wives' tales and the new world system: Gilbert, Galileo and Kepler', *Configurations*, **7** (3), 301–57.
34. Rosen (1967), p. 14.
35. Taussig, M. (1993), *Mimesis and Alterity: A Particular History of the Senses*, London and New York: Routledge.
36. Ibid., p. 25.
37. Taussig says, for instance, of the magical 'copy' of Western ethnographer Stephanie Kane (not a user of the 'ethnographic present', but a narrator of singular events): 'Kane's mode relies not on abstract general locutions such as "among the Emberá it is believed that …", but instead concentrates on image-ful particularity in such a way that … she creates like magical reproduction itself, a sensuous sense of the real, mimetically at one with what it attempts to represent. … [C]an't we say that *to give an example, to instantiate, to be concrete*, are all examples of the magic of mimesis wherein the replication, the copy, acquires the power of the represented?' (1993, p. 16).
38. Ibid., p. 25.
39. It is perhaps slightly more than a strange irony that the people of Kepler and his mother's home town leapt so almost fatally to the conclusion that this mimetic mother was Kepler's mother, and that her death really did put this text into circulation. At any rate, it wouldn't have surprised the Cuna.
40. In this, his book reads very like Thomas Harriot's *Virginia*, which also saves its concise remarks on the inhabitants and their customs until the very end: Harriot, T. (1590), *A Briefe and True Report of the New Found Land of Virginia*, Frankfurt: Theodore De Bry. Pratt (1993) has definitively explained this tendency in earthly, or at least colonialist, voyage-writing: it is, after all, land and the climate that

European governments and settlers want, and a depopulated description permits, at some level, depopulation. Or, at least, the extreme marginalisation of any indigenous competition in the homebound reader's imagined landscape. Here Kepler parts ways with his fellow voyagers.

41. Rosen (1967), p. 21.
42. Ibid., p. 24.
43. Jameson, F.R. (1981), *The Political Unconscious: Narrative as a Socially Symbolic Act*, Ithaca: Cornell University Press, p. 112.
44. See Kepler's Note 164 (and Note 169): 'because the moon moves around the earth in the same direction as the earth's surface around its axis, the … lower hemisphere of the earth or Volva seems to the moondwellers to be travelling westward' (Rosen, 1967, p. 114).
45. See Kenseth, J. (1991), 'A World of Wonders in One Closet Shut', in J. Kenseth (ed.), *The Age of the Marvelous*, Hanover, NH: Hood Museum of Art, pp. 81–102, and Evans, R.J.W. (1997), *Rudolf II and His World: A Study in Intellectual History, 1576–1612*, London: Thames and Hudson.
46. Rosen (1967), p. 28.
47. For more on the amazing Kircher, see Allen, D.C. (1949), *The Legend of Noah: Renaissance Rationalism in Art, Science, and Literature*, Urbana, IL: University of Illinois Press; and Findlen, P. (1994), *Possessing Nature: Museums, Collecting, and Scientific Culture in Early Modern Europe*, Berkeley, CA: University of California Press.
48. Rosen (1967), p. 27.
49. Ibid., p. 151.
50. Ibid., p. 28.

Select Bibliography

Acosta, J. ([1604] 1880), *The Natural and Moral History of the Indies*, trans. E. Grimston, ed. C.R. Markham, 2 vols, London: Hakluyt Society.

Albanese, D. (1996), *New Science, New World*, Durham, NC: Duke University Press.

Alexander, M. (ed.) (1976), *Discovering the New World, Based on the Works of Theodore De Bry*, London: London Editions.

Allen, D.C. (1949), *The Legend of Noah: Renaissance Rationalism in Art, Science, and Literature*, Urbana, IL: University of Illinois Press.

Appleby, J.H. (1986), 'Archibald Pitcairne Re-Encountered: A Note on his Manuscript Poems and Printed Library Catalogue', *The Biblioteck*, **12** (6), 137–9.

Aristotle (1958), *Generation of Animals*, trans. A.L. Peck, Cambridge and London: Harvard University Press.

—— (1984), *The Complete Works of Aristotle*, ed. J. Barnes, 2 vols, Guildford: Princeton University Press.

Arons, W. (ed. and trans.) (1994), *When Midwifery Became the Male Physician's Province: the Sixteenth Century Handbook 'The Rose Garden for Pregnant Women and Midwives', Newly Englished*, Jefferson, NC: McFarland.

Bacon, F. (1857–74), *The Works of Francis Bacon*, 14 vols, ed. J. Spedding, R.L. Ellis and D.D. Heath, London: Longman.

Bakhtin, M. (1968), *Rabelais and His World*, trans. H. Iswolsky, Cambridge, MA: MIT Press.

Bald, R.C. (1970), *John Donne: A Life*, Oxford: Clarendon Press.

Bartholin, T. (1668), *Bartholinus Anatomy*, trans. N. Culpeper and A. Cole, London: printed by John Streeter.

Baudrillard, J. (1995), *Simulacra and Simulation*, trans. S. Glaser, Ann Arbor, MI: University of Michigan Press.

Berengario da Carpi, J. (1521), *Carpi Commentaria cum amplissimus Additionibus super Anatomia Mundini una cum textu eiusdem in pristinum et verum nitorem redacto*, Bologna: n.p.

Bhabha, H. (1994), 'Articulating the Archaic: Cultural difference and colonial nonsense', in Bhabha, H., *The Location of Culture*, London: Routledge, pp. 123–38.

Biagioli, M. (1993), *Galileo, Courtier: The Practice of Science in the Culture of Absolutism*, Chicago: University of Chicago Press.

Blumenfeld-Kosinski, R. (1990), *Not of Woman Born: Representations of Caesarean Birth in Medieval and Renaissance Culture*, Ithaca and London: Cornell University Press.

Bono, J. (1995), *The Word of God and the Languages of Man: Interpreting Nature in Early Modern Science and Medicine*, Madison, WI: University of Wisconsin Press.

Bordo, S. (1986), 'The Cartesian Masculinization of Thought', *Signs: Journal of Women in Culture and Society*, **11**, 439–56.

Boucé, P.-G. (1987), 'Imagination, pregnant women, and monsters, in eighteenth-century England and France', in G.S. Rousseau and R. Porter (eds), *Sexual Underworlds of the Enlightenment*, Manchester: Manchester University Press, pp. 86–100.

Bower, A. (1817), *The History of the University of Edinburgh*, 2 vols, Edinburgh and London: Oliphant, Waugh, Innes and Murray.

Boyle, R. (1660), *New Experiments Physico-Mechanical, Touching the spring of the Air and its effects*, Oxford: H. Hall for T. Robinson.

Bradstock, A. (1997), *Faith in the Revolution: The political theologies of Müntzer and Winstanley*, London: SPCK.

—— (ed.) (2000), *Winstanley and the Diggers 1649–1999*, London and Portland, OR: Frank Cass.

Brodie, A. (ed.) (1997), *The Scottish Enlightenment: an Anthology*, Edinburgh: Canongate.

Brown, P. (1985), '"This thing of darkness I acknowledge mine": *The Tempest* and the discourse of colonialism', in J. Dollimore and A. Sinfield (eds), *Political Shakespeare*, Manchester: Manchester University Press, pp. 48–71.

Browne, P. (ed.) (1999), *Reading Dreams: The Interpretation of Dreams from Chaucer to Shakespeare*, Oxford: Oxford University Press.

Browne, T. (1967), *The Prose of Sir Thomas Browne*, ed. N.J. Endicott, New York: Anchor Books.

—— (1977), *The Major Works*, ed. C.A. Patrides, Harmondsworth, Penguin.

Calvi, G. (1988), 'The Florentine Plague of 1630–33: Social Behaviour and Symbolic Action', in N. Bulst and R. Delort (eds), *Maladies et Société (XIIe–XVIIIe siècles)*, Paris: Centre National de la Recherche Scientifique, pp. 327–36.

Camporesi, P. (1988), *The Incorruptible Flesh: Bodily Mutation and Mortification in Religion and Folklore*, trans. T. Croft-Murray, Latin texts trans. H. Elsom, Cambridge: Cambridge University Press.

Capkova, D. (1994), 'Comenius and his Ideals: Escape from the Labyrinth', in M. Greengrass, M. Leslie and T. Raylor (eds), *Samuel Hartlib and Universal Reformation*, Cambridge: Cambridge University Press.

Cassirer, E. (1963), *The Individual and the Cosmos in Renaissance Philosophy*, trans. M. Domandi, Oxford: Basil Blackwell.

Cavendish, M. (1653), *Poems and Fancies*, London: T.R. for J. Martin & J. Allestrye.
—— (1655), *The Philosophical and Physical Opinions*, London: F. Martin and F. Allestrye.
—— (1655), *The World's Olio*, London: J. Martin and J. Allestrye.
—— (1666), *Observations Upon Experimental Philosophy*, London: F. Martin and F. Alleystre.
—— (1668), *Grounds of Natural Philosophy*, London: A. Maxwell.
—— (1994), *The Blazing World and other writings*, ed. K. Lilley, London: Penguin.
Chamberlen, H. (trans.) (1672), *The Diseases of Women with Child*, London: R Clavel, W. Cooper and Benjamin Billingsly.
—— (trans.) (1673), *The Accomplisht Midwife, Treating of the Diseases of Women with Child*, London: Benjamin Billingsley.
Chambers, A.B. (1963), '"Sin" and "Sign" in *Paradise Lost*', *Huntingdon Library Quarterly*, **26**, 381–2.
Choulant, L. (1962), *History and Bibliography of Anatomic Illustration*, trans. M. Frank, London: Hafner.
Christianson, G.E. (1976), 'Kepler's *Somnium*: Science fiction and the renaissance scientist', *Science Fiction Studies*, **3** (2), 79–90.
Clapham, H. (1603), *An Epistle discovrsing vpon the present Pestilence*, London: Printed by T.C. for the Widow Newbery.
Clucas, S. (1994), 'In Search of "The True Logick": Methodological Eclecticism among the "Baconian Reformers"', in M. Greengrass, M. Leslie and T. Raylor (eds), *Samuel Hartlib and Universal Reformation*, Cambridge: Cambridge University Press, pp. 51–74.
Colie, R. (1966), *Paradoxia Epidemica: The Renaissance Tradition of Paradox*, Princeton: Princeton University Press.
Cormack, L. (1997), *Charting an Empire: Geography at the English Universities, 1580–1620*, Chicago: University of Chicago Press.
Cosgrove, D. (1992), *The Palladian Landscape*, Leicester: Leicester University Press.
—— (1999), *Mappings*, London: Reaktion.
Cowley, A. (1905), *Abraham Cowley: Poems*, ed. A.R. Waller, Cambridge: Cambridge University Press.
Crawford, P. (1981), 'Attitudes to Menstruation in Seventeenth-Century England', *Past and Present*, **91**, 47–73.
—— (1990), 'The construction and experience of maternity in seventeenth-century England', in V. Fildes (ed.), *Women as Mothers in Pre-Industrial England: Essays in Memory of Dorothy McLaren*, London and New York: Routledge, pp. 3–38.

Crooke, H. (1651), *Microcosmographia: a Decription of the Body of Man*, London: n.p.

Culpeper, N. (1651), *A Directory for Midwives*, London: printed by Peter Cole.

Cunningham, A. (1981), 'Sydenham versus Newton: the Edinburgh Fever Dispute in the 1690s', in W.F. Bynum and V. Nutton (eds), *Theories of Fever from Antiquity to the Enlightenment*, London: Wellcome Institute.

—— (1997), *The Anatomical Renaissance: The Resurrection of the Anatomical Projects of the Ancients*, Aldershot: Scolar Press, Ashgate Publishing.

Dally, A. (1991), *Women Under the Knife: a History of Surgery*, London: Hutchinson.

Daston, L. (1998), *Wonders and the Order of Nature, 1150–1750*, London: Zone Books.

Davis, N.Z. (1975), *Society and Culture in Early Modern France*, Stanford, CA: Stanford University Press.

Debus, A.G. (1978), *Man and Nature in the Renaissance*, Cambridge: Cambridge University Press.

Dee, J. (1650), 'Mathematicall Preface', in Euclid, *Euclides Elementes of Geometry: The First VI Books*, trans. T. Rudd, London: n.p.

De Foigny, G. (1693), *A New Discovery of Terra Incognita Australi, or the Southern World*, London: n.p.

Descartes, R. (1968), *Descartes: Discourse on Method and the Mediations*, trans. and ed. F.E. Sutcliffe, Harmondsworth: Penguin.

—— (1981), *Descartes: Philosophical Letters*, trans. and ed. A. Kenny, Oxford: Basil Blackwell.

—— (1985), *Philosophical Writings*, trans. J. Cottingham, R. Stoothof and D. Murdoch, 3 vols, Cambridge: Cambridge University Press.

Dick, Steven J. (1982), *Plurality of Worlds: The Origins of the Extraterrestrial Life Debate from Democritus to Kant*, Cambridge: Cambridge University Press.

Dingwall, H.M. (1995), *Phyisicans, Surgeons and Apothocaries: Medicine in Seventeenth-Century Edinburgh*, East Linton: Tuckwell Press.

Donegan, J.B. (1978), *Women and Men Midwives: Medicine, Morality, and Misogyny in Early America*, Westport, CT and London: Greenwood Press.

Donne, J. (1930), *Biathanatos*, New York: Facsimile Text Society.

—— (1952), *John Donne: Essayes in Divinity*, ed. E.M. Simpson, Oxford: Clarendon Press.

—— (1953–62), *The Complete Sermons of John Donne*, ed. G.R. Potter and E.M. Simpson, 10 vols, Berkeley and Los Angeles, CA: California University Press.

—— (1967), *Satires, Epigrams and Verse Letters*, ed. W. Milgate, Oxford: Clarendon Press.

—— (1969), *Ignatius His Conclave*, ed. T.S. Healy, Oxford: Clarendon Press.
—— (1975), *The Complete English Poems*, ed. A.J. Smith, Harmondsworth: Penguin.
—— (1980), *John Donne: Paradoxes and Problems*, ed. H. Peters, Oxford: Clarendon Press.
—— (1985), *Complete English Poems*, ed. C.A. Patrides, London: J.M. Dent & Sons.
Donnison, J. (1977), *Midwives and Medical Men: A History of Inter-Professional Rivalries and Women's Rights*, London: Heinemann.
Du Bartas, G. (1979), *The Divine weeks and Works of Guillaume de Saluste Sieur Du Bartas*, trans. J. Sylvester, ed. S. Snyder, 2 vols, Oxford: Clarendon Press.
Duncan, D. (1965), *Thomas Ruddiman: a Study in Scottish Scholarship of the early Eighteenth-Century*, Edinburgh and London: Oliver and Boyd.
—— (1987), 'Scholarship and Politeness in the Early Eighteenth Century', in A. Hooke (ed.), *The History of Scottish Literature: 1660–1800*, gen. ed. C. Craig, 4 vols, Aberdeen: Aberdeen University Press, vol. 2, pp. 51–63.
Eccles, A. (1982), *Obstetrics and Gynaecology in Tudor and Stuart England*, London and Canberra: Croom Helm.
Edgerton, S.Y., Jr (1985), *Pictures and Punishment: Art and Criminal Prosecution during the Florentine Renaissance*, Ithaca and London: Cornell University Press.
Edwards, K. (1999), *Milton and the Natural World: Science and Poetry in Paradise Lost*, Cambridge: Cambridge University Press.
Emerson, R.L. (1988a) 'Science and the Origins and Concerns of the Scottish Enlightenment', *The History of Science*, **26**, 333–65.
—— (1988b) 'Sir Robert Sibbald, Kt, the Royal Society of Scotland and the Origins of the Scottish Enlightenment', *Annals of Science*, **45**, 41–72.
—— (1989), 'The religious, the secular and the worldly: Scotland 1680–1800', in J. Crimmins (ed.), *Religion, Secularization and Political Thought*, London and New York: Routledge, pp. 68–89.
Erickson, R.A. (1982), '"The Books of Generation": Some Observations on the Style of the British Midwife Book, 1671–1764', in Boucé, P.-G. (ed.), *Sexuality in Eighteenth-Century Britain*, Manchester: Manchester University Press and Totowa: Barnes and Noble, pp. 74–94.
—— (1997), *The Language of the Heart 1600–1750*, Philadelphia: University of Pennsylvania Press.
Estienne, C. (1545), *De Dissectione Partium Corporis Humani libri tres*, Paris: n.p.
Evenden-Nagy, D. (1991), 'Seventeenth Century London Midwives: Their Training, Licensing and Social Profile', unpublished PhD dissertation, McMaster University.

Farrington, B. (1951), '*Temporis Partus Masculus*: An Untranslated Writing of Francis Bacon', *Centaurus*, **1**, 193–205.

Feingold, M. (1984), *The Mathematician's Apprenticeship: Science, Universities and Society in England, 1560–1640*, Cambridge: Cambridge University Press.

Fildes, V. (1986), *Breasts, Bottles and Babies: A History of Infant Feeding*, Edinburgh: Edinburgh University Press.

—— (1988), *Wet Nursing: A History from Antiquity to the Present*, Oxford: Basil Blackwell.

Findlen, P. (1994), *Possessing Nature: Museums, Collecting, and Scientific Culture in Early Modern Italy*, Berkeley, CA: University of Berkeley Press.

Fish, S. (1967), *Surprised by Sin: The Reader in Paradise Lost*, London: Macmillan.

Fissell, M. (1995), 'Gender and Generation: Representing Reproduction in Early Modern England', *Gender and History*, **7** (3), 433–56.

Foucault, M. (1970), *The Order of Things: An Archaeology of the Human Sciences*, trans. from the French, London: Routledge.

Fradenburg, L. and C. Freccero (eds) (1996), *Premodern Sexualities*, New York: Routledge.

French, P. (1972), *John Dee: The World of an Elizabethan Magus*, New York and London: Routledge.

French, R. (1994), *William Harvey's Natural Philosophy*, Cambridge: Cambridge University Press.

—— (1999), *Dissection and Vivisection in the European Renaissance*. Aldershot, UK and Brookfield, VT: Ashgate.

Fuller, M. (1995), *Voyages in Print: English Travel to America, 1576–1624*, Cambridge: Cambridge University Press.

Gallagher, P.J. (1976), '"Real or Allegoric": the Ontology of Sin and Death in *Paradise Lost*', *English Literary Renaissance*, **6**, 317–35.

Godwin, F.B. (1638), *The Man in the Moon*, trans. E.M., London: n.p.

Goldberg, J. (ed.) (1994), *Queering the Renaissance*, Durham, NC and London, Duke University Press.

Gosse, E. (1899), *The Life and Letters of John Donne*, 2 vols, London: William Heinemann.

Grant, D. (1957), *Margaret the First*, London: Rupert Hart-Davis.

Greenblatt, S. (1988), *Shakespearean Negotiations: The Circulation of Social Energy in Renaissance England*, Oxford: Clarendon Press.

—— (1991), *Marvelous Possessions: The Wonder of the New World*, Oxford: Clarendon Press.

Greengrass, M., Leslie, M. and Raylor, T. (eds) (1994), *Samuel Hartlib and the Universal Reformation*, Cambridge: Cambridge University Press.

Groneman, C. (1995), 'Nymphomania. The Historical Construction of Female

Sexuality', in J. Terry and J. Urla (eds), *Deviant Bodies. Critical Perspectives on Difference in Science and Popular Culture*, Bloomington and Indianapolis, IN: Indiana University Press, pp. 219–49.

Guerrini, A. (1985), 'James Keill, George Cheyne, and Newtonian Physiology', *1690–1740, Journal of the History of Biology*, **18** (12), 247–66.

—— (1987), 'The Tory Newtonians: Gregory, Pitcairne and their Circle', *Journal of British Studies*, **25**, 288–311.

—— (1987), 'Archibald Pitcairne and Newtonian Medicine', *Medical History*, **31**, 70–93.

—— (1993), '"A Club of Little Villians": rhetoric, professional identity and medical pamphlet wars' in M. Roberts, M. Porter and R. Porter (eds), *Literature and Medicine During the Eighteenth Century*, London and New York: Routledge, pp. 226–44.

—— (2000), *Obesity and Depression in the Enlightenment: The Life and Times of George Cheyne*, Norman, OK: Oklahoma University Press.

Guillemeau, J. (1612), *Childbirth Or, The Happy Deliverie of Women*, London: A. Hadfield.

Haber, F.C. (1986), 'Time, Technology, Religion, and Productivity Values in Early Modern Europe', in F.C. Faber, *Time, Science, and Society in China and the West*, Amherst, MA: University of Massachusetts Press, pp. 79–92.

Hadfield, A. (1998), *Literature, Travel and Colonial Writing in the English Renaissance, 1545–1625*, Oxford: Clarendon Press.

Hakluyt, R. ([1600] 1903), *The Principal Navigations, Voyages, Traffiques & Discoveries of the English Nation*, 12 vols, Glasgow: MacLehose.

Hall, A.R. (1963), *From Galileo to Newton*, London: Collins.

Hall, J. (1605), *Mundus alter et idem*, London: n.p.

Hanke, L. (1959), *Aristotle and the American Indian: A Study in Race Prejudice in the Modern World*, London: Hollis & Carter.

Hartlib, S. (1655), *The Reformed Commonwealth of Bees. Presented in Several Letters and Observations*, London: Giles Calvert.

Harvey, W. (1653), *Anatomical Exercitations, Concerning the Generation of Living Creatures*, London: n.p.

—— (1990), *The Circulation of the Blood and Other Writings*, trans. K.J. Franklin, intro. A. Wear, London and Rutland, Vermont: Everyman.

Haynes, R.D. (1994), *From Faust to Strangelove: Representations of the Scientist in Western Literature*, Baltimore, MD: Johns Hopkins University Press.

Heath, T. (1949), *Mathematics in Aristotle*, Oxford: Clarendon Press.

Heckscher, W.S. (1958), *Rembrandt's Anatomy of Dr. Nicolaas Tulp: An Iconological Study*, New York: New York University Press.

Helgerson, R. (1992), *Forms of Nationhood: The Elizabethan Writing of England*, Chicago: University of Chicago Press.

Henderson, J. (1988), 'Epidemics in Renaissance Florence: Medical Theory and Government Response', in N. Bulst and R. Delort (eds), *Maladies et Société (XIIe–XVIIIe siècles)*, Paris: Centre National de la Recherche Scientifique, pp. 165–86.

Hess, A.G. (1994), 'Community Case Studies of Midwives from England and New England, *c.* 1650–1720', unpublished PhD thesis, University of Cambridge.

Hill, C. (ed.) (1973), *Winstanley: The Law of Freedom and Other Writings*, Harmondsworth: Penguin.

—— (1978), *The Religion of Gerrard Winstanley*, Oxford: The Past and Present Society.

Hobbes, T. (1991), *Leviathan*, ed. R. Tuck, Cambridge: Cambridge University Press.

Hobby, E. (1988), *Virtue of Necessity: English Women's Writing 1649–1688*, London: Virago, Ann Arbor, MI: Michigan University Press.

—— (2001a), '"Delight in a singularity": Margaret Cavendish, Duchess of Newcastle, in 1671', *In-between: Essays and Studies in Literary Criticism.*

—— (2003), 'Yarhound, Horrion, and the horse-headed Tartar: editing Jane Sharp's *The Midwives Book* (1671)', in K. Binhammer and J. Wood (eds), *Feminist Literary History Re(Dis)Covered*, Newark, DE: Delaware University Press.

Hogden, M.T. (1964), *Early Anthropology in the Sixteenth and Seventeenth Centuries*, Philadelphia: University of Pennsylvania Press.

Holstun, J. (2000), *Ehud's Dagger: Class Struggle in the English Revolution*, London and New York: Verso.

Hood, T. (1974), *A Copie of the Speache Made by The Mathematicall Lecturer ... at the House of Thomas Smith*, Amsterdam: Theatrum Orbis Terrarum.

Hooke, R. ([1665] 1961), *Micrographia*, preface R.T. Gunther, New York: Dover.

Huet, M.-H. (1993), *Monstrous Imagination*, Cambridge, MA: Harvard University Press.

Hulme, P. (1986), *Colonial Encounters: Europe and the Native Caribbean, 1492–1797*, London: Methuen.

Hulton, P. (ed.) (1984), *America 1585: The Complete Drawings of John White*, London: British Museum.

—— and Quinn, D.B. (eds) (1964), *The American Drawings of John White, 1577–1590*, 2 vols, London: British Museum.

Hunter, M. (1981), *Science and Society in Restoration England*, Cambridge: Cambridge University Press.

—— (1992), '"Aikenhead the Atheist": The Context and Consequences of Articulate Irreligion in the Late Seventeenth Century', in M. Hunter and

D. Wooton (eds), *Atheism from the Reformation to the Enlightenment*, Oxford: Clarendon Press, pp. 221–54.

—— (1995), *Science and the Shape of Orthodoxy: Intellectual Change in Late Seventeenth-Century Britain*, Woodbridge: Boydell.

Ingerson, A.E. (1994), 'Tracking and testing the nature/culture dichotomy in practice', in C. Crumley (ed.), *Historical Ecology: Cultural Knowledge and Changing Landscapes*, Santa Fe: SAR Press, pp. 43–66.

Irving, D. (1839), *Lives of the Scottish Writers*, Edinburgh: n.p.

Jacob, G. (1718), *Tractatus de Hermaphroditis: Or, a Treatise of Hermaphrodites*, London: n.p.

Jameson, F.R. (1981), *The Political Unconscious: Narrative as a Socially Symbolic Act*, Ithaca: Cornell University Press.

Jardine, L. (1996), *Worldly Goods*, London: Macmillan.

Jay, M. (1994), *Downcast Eyes*, London: University of California Press.

Jenner, M.S.R. (1999), 'Body, Image, Text in Early Modern Europe', *Social History of Medicine*, **12**, 143–54.

Jinner, S. (1658), *An Almanack or Prognostication for the year of our Lord 1658*, London: Stationers' Company.

—— (1659), *An Almanack or Prognostication for the year of our Lord 1659*, London: Stationers' Company.

—— (1660), *An Almanack or Prognostication for the year of our Lord 1660*, London: Stationers' Company.

—— (1664), *An Almanack for the year of our Lord God 1664*, London: Stationers' Company.

Johnson, S. (1947), 'John Donne and the Virginia Company', *English Literary History*, **14**, 127–38.

Johnston, W.T. (ed.) (1979), *The Best of our Owne: The Letters of Archibald Pitcairne, 1652–1713*, Edinburgh: Saorsa.

Jollat, Mercure (1557), *Les Figures et Portraicts des Parties du Corps Humain*, Paris: J. Kerver.

—— (1575), *Les Figures et Portraicts des Parties du Corps Humain*, Paris: J. Kerver.

Jonas, R. (trans.) (1540), *The byrth of mankynd*, London: T.R.

Jones, K. (1988), *A Glorious Fame: The Life of Margaret Cavendish, Duchess of Newcastle, 1623–1673*, London: Bloomsbury.

Jonson, B. (1975), *Poems*, ed. I. Donaldson, Oxford: Oxford University Press.

Jowitt, C. (1995), 'Radical Identities? Native Americans, Jews and the English Commonwealth', *The Seventeenth Century*, **10**, 101–19.

Kaplan, R.L. (2000), *The Nothing That Is: A Natural History of Zero*, New York: Oxford University Press.

Keller, E.F. (1980), 'Baconian Science: A Hermaphroditic Birth', *The Philosophical Forum*, **3**, 299–308.

—— (1985), *Reflections on Gender and Science*, New Haven, CT: Yale University Press.
—— (1992), *Secrets of Life, Secrets of Death: Essays on Language, Gender and Science*, New York and London: Routledge.
—— (1995), 'Mrs Jane Sharp: Midwifery and the Critique of Medical Knowledge in Seventeenth-century England', *Women's Writing: the Elizabethan to Victorian Period*, **2** (2), 101–11.
Kellett, C.E. (1957), 'A Note on Rosso and the Illustrations to Charles Estienne's *De dissectione*', *Journal of the History of Medicine*, **12**, 325–36.
Kepler, J. (1634), *Somnium, seu opus posthumous de astronomi lunari*, Zagan and Frankfurt-am-Main: n.p.
Keynes, G. (1973), *A Bibliography of Dr. John Donne*, Oxford: Clarendon Press.
Kidd, C. (1999), *British Identities Before Nationalism: Ethnicity and Nationhood in the Atlantic World, 1600–1800*, Cambridge: Cambridge University Press.
Kiple, K.F. (ed.) (1993), *The Cambridge World History of Human Disease*, Cambridge: Cambridge University Press.
Knapp, J. (1992), *An Empire Nowhere: England, America and Literature from Utopia to the Tempest*, Berkeley, CA: University of California Press.
Kristeva, J. (1982), *Powers of Horror: An Essay on Abjection*, trans. L.S. Roudiez, New York: Columbia University Press.
Kuhn, T.S. (1957), *The Copernican Revolution*, Cambridge, MA and London: Harvard University Press.
Kuppermann, K.O. (1980), *Settling with the Indians: The meeting of English and Indian cultures in America, 1580–1640*, Totowa, NJ: Rowman & Allanheld.
Laqueur, T. (1990), *Making Sex: Body and Gender from the Greeks to Freud*, Cambridge, MA and London: Harvard University Press.
Lear, J. (ed.) (1965), *Kepler's Dream*, trans. P.F. Kirkwood, Berkeley and Los Angeles: University of California Press.
Lewalski, B. (1994), 'Milton and the Hartlib Circle: Educational Projects and Epic *Paideia*', in D.T. Benet and M. Lieb (eds), *Literary Milton: Text, Pretext, Context*, Pittsburgh, PA: Duquesne University Press, pp. 202–19.
Lim, W.S.H. (1998), *The Arts of Empire: The Poetics of Colonialism From Ralegh to Milton*, London: Associated University Press.
Lind, L.R. (1949), *The Epitome of Andreas Vesalius*, New York: Macmillan.
Linden, S.J. (1996), *Dark Hierogliphicks: Alchemy in English Literature from Chaucer to the Restoration*, Lexington, KT: University of Kentucky Press.
Lindenboom, G.A. (1963), 'Pitcairne's Leyden Interlude Described from the Documents', *Annals of Science*, **19**, 273–84.
Linton, J.P. (1998), *The Romance of the New World: Gender and the Literary*

Formations of English Colonialism, Cambridge: Cambridge University Press.
Loomba, A. (1996), 'Shakespeare and Cultural Difference', in T. Hawkes (ed.), *Alternative Shakespeares*, 2 vols, London and New York: Routledge, vol. 2, pp. 164–91.
Lucian (1927), *A True Story (Verae Historiae)*, trans. A.M. Harmon, 8 vols, London: William Heinemann.
Mackenzie, G. (1705), *A Bundle of Papers. Partly Self-Evident, partly Problematick raised from Occasional Meditations*, Edinburgh: n.p.
Macqueen, J. (1982), *Progress and Poetry: The Enlightenment and Scottish Literature*, Edinburgh: Scottish Academic Press.
Maltby, W.S. (1971), *The Black Legend in England: The Development of Anti-Spanish Sentiment, 1558–1660*, Durham, NC: Duke University Press.
Marland, H. (ed.) (1993), *The Art of Midwifery: Early Modern Midwives in Europe*, London and New York: Methuen.
Martensen, R. (1994), 'The Transformation of Eve: Women's Bodies, Medicine and Culture in Early Modern England', in R. Porter and M. Teich (eds), *Sexual Knowledge, Sexual Science*, Cambridge and New York: Cambridge University Press, pp. 107–33.
McRae, A. (1992), 'Husbandry Manuals and the Language of Agrarian Improvement', in M. Leslie and T. Raylor (eds), *Culture and Cultivation in Early Modern England: Writing and the Land*, London: Leicester University Press, pp. 35–62.
Merchant, C. (1980), *The Death of Nature: Women, Ecology and the Scientific Revolution*, San Francisco: Harper & Row/Harper.
Merlan, P. (1960), *From Platonism to Neoplatonism*, The Hague: Nijhoff.
Milton, J. (1953), *Complete Prose Works*, ed. D.M. Wolfe, New Haven, CT: Yale University Press.
—— (1966), *Poetical Works*, ed. D. Bush, Oxford: Oxford University Press.
—— ([1667] 1971), *Paradise Lost*, ed. A. Fowler, London: Longman.
Mintz, S.I. (1952), 'The Duchess of Newcastle's visit to the Royal Society', *Journal of English and Germanic Philology*, **51**, 165–76.
Molins, W. (1648), *Muskotomia: or, the Anatomical Administration of all the Muscles*, London: Edward Husband.
Montaigne, M. de (1580 ff), *Essais*, Bordeaux.
—— ([1603] 1910), 'Of the Cannibals', in *Essayes*, trans. J. Florio, 3 vols, London: Everyman, vol. 1, pp. 215–29.
More, Sir Thomas (1516), *Utopia*, Louvain: n.p.
Mulder, D. (1990), *The Alchemy of Revolution: Gerrard Winstanley's Occultism and Seventeenth-Century English Communism*, New York: Peter Lang.
Nerlich, M. (1987), *Ideology of Adventure: Studies in Modern Consciousness,*

1100–1750, vol. 1, trans. R. Crowley, Minneapolis, MN: University of Minnesota Press.

Neville, H. (1668), *The Isle of Pines*, London: n.p.

Nicolson, M.H. (ed.) (1930), *Conway Letters: the Correspondence of Anne, Viscountess Conway, Henry More, and their Friends, 1642–1684*, London: Oxford University Press.

—— (1966), 'The Discovery of Space', in O.B. Hardison (ed.), *Medieval and Renaissance Studies: Proceedings of the Southeast Institute for Renaissance Studies*, Chapel Hill, NC: University of North Carolina Press, pp. 40–59.

Nuti, L. (1988), 'The Mapped Views by Georg Hoefnagel: The Merchant's Eye, the Humanist's Eye', *Word and Image*, **4** (2), 545–70.

Oliphant, C. (1702), *A Short Answer to Two Lybels Lately Published*, Edinburgh: n.p.

Orkin, M. (1997), 'Whose things of darkness?', in J.J. Joughin (ed.), *Shakespeare and National Culture*, Manchester: Manchester University Press, pp. 142–69.

Ouston, H. (1982), 'York in Edinburgh, James VII and the Patronage of Learning in Scotland 1676–1688', in J. Dwyer, Roger A. Mason and A. Murdoch (eds), *New Perspectives on the Politics and Culture of Early Modern Scotland*, Edinburgh: John Donald, pp. 133–55.

Pagden, A. (1982), *The Fall of Natural Man: The American Indian and the Origins of Comparative Ethnology*, Cambridge: Cambridge University Press.

—— (1993), *European Encounters with the New World*, New Haven, CT: Yale University Press.

Paré, A. (1678), *The Works of that famous chirurgeon Ambrose Parey*, trans. T. Johnson, London: Printed by Mary Clark for John Clark.

—— ([1573] 1982), *On Monsters and Marvels*, trans. J.L. Pallister, Chicago: Chicago University Press.

Park, K. (1994), 'The Criminal and the Saintly Body: Autopsy and Dissection in Renaissance Italy', *Renaissance Quarterly*, **47**, 1–33.

—— and Daston, L. (eds), *The Cambridge History of Science, Early Modern Science Volume 3*, Cambridge: Cambridge University Press (forthcoming 2003).

—— and Nye, R.A. (1991), 'Destiny is Anatomy', *The New Republic*, 18 February, 53–7.

Parker, M. ([1637] 1927), 'The Two Inseparable Brothers, or a Strange Description of a Gentleman Who hath an Imperfect Brother Growing out of His Side', in H.E. Rollins (ed.), *The Pack of Autolycus*, Cambridge, MA: Harvard University Press, pp. 7–14.

Paster, G.K. (1987), 'Leaky Vessels: the Incontinent Women of City Comedy', *Renaissance Drama*, **18**, 43–65.

Perkins, W. (1996), *Midwifery and Medicine in Early Modern France*, Exeter: Exeter University Press.

Phaire, T. (1993), *The Regiment of Life, whereunto is added a treatise of the Pestilence* (1553), transcribed and intr. A.V. Neale and H.R.E. Wallis, London: British Paediatric Association.

Phillipson, N. (1989), *Hume*, London: Weidenfield and Nicholson.

Pitcairne, A. (1715), *The Works*, trans. G. Sewell and J.T. Desaguliers, London: Curll.

—— (1727) *Selecta Poemata*, Edinburgh: Freebairn.

—— (1830), *Babell: a Satirical Poem on the Proceedings of the General Assembly in the Year M.DC.XCII*, Edinburgh: Maitland Club.

Plattes, G. ([1641] 1970), *A Description of the Famous Kingdom of Macaria*, in Webster, C. (ed.), *Samuel Hartlib and the Advancement of Learning*, Cambridge: Cambridge University Press, pp. 79–90.

Pocock, J.G.A. (1970), 'Time, History and Eschatology in the Thought of Thomas Hobbes', in J.H. Elliott and H.G. Koenigsberger (eds), *The Diversity of History: Essays in Honour of Sir Herbert Butterfield*, London: Routledge & Kegan Paul, pp. 149–98.

Porter, R. (1987), *Disease, Medicine and Society in England 1550–1860*, London: Macmillan.

—— (1991), 'Introduction', in S. Pumfrey, Paolo L. Rossi and Maurice Slawinski (eds), *Science, Culture and Popular Belief in Renaissance Europe*, Manchester: Manchester University Press, pp. 1–12.

—— (1997), *The Greatest Benefit to Mankind: A Medical History of Humanity from Antiquity to the Present*, London: Harpercollins.

Power, D. (1927), 'The Birth of Mankind or the Woman's Book: a Bibliographical Study', *The Library*, **8** (1), 1–37.

Pratt, M.L. (1993), *Imperial Eyes: Travel Writing and Transculturation*, London and New York: Routledge.

Price, B. (1996), 'Feminine modes of knowing and scientific enquiry: Margaret Cavendish's poetry as case study', in H. Wilcox (ed.), *Women and Literature in Britain 1500–1700*, Cambridge: Cambridge University Press.

Proclus (1970), *A Commentary on the First Book of Euclid's Elements*, trans. and ed. G.R. Morrow, Princeton, NJ: Princeton University Press.

Purver, M. (1967), *The Royal Society: Concept and Creation*, London: Routledge and Kegan Paul.

Quinn, D.B. (ed.) (1955), *The Roanoke Voyages, 1584–1590*, 2 vols, London: Hakluyt Society.

—— (ed.) (1974), *The Hakluyt Handbook*, 2 vols, London: Hakluyt Society.

R.C., I.D., M.S., T.B. (1659), *The Compleat Midwife's Practice Enlarged*, London: N. Brook.

Rahner, K. (1961), *On the Theology of Death* [*Quaestiones Disputatae*], trans. C.H. Henkey, Freiburg: Herder; Edinburgh and London: Nelson.

Ralegh, W. (1596), *The Discoverie of the Large, Rich and Bewtiful Empyre of Guiana*, London: n.p.

de la Ramée, P. (1569), *Scholae Mathematicae*, London: n.p.

—— (1590), *The Elementes of Geometrie. Written in Latin by That Excellent Scholler P. Ramus*, trans. T. Hood, London: n.p.

Raynold, T. (trans.) (1545), *The Byrth of Mankynd*, London: T. Ray.

Reeves, E. (1999), 'Old wives' tales and the new world system: Gilbert, Galileo and Kepler', *Configurations*, **7** (3), 301–57.

Reiss, T. (1982), *The Discourse of Modernism*, Ithaca: Cornell University Press.

Riley, P.W.J. (1979), *King William and the Scottish Politicians*, Edinburgh: John Donald.

Riolan, J. (1671), *A Sure Guide*, trans. N. Culpeper, 3rd edn, London: George Sawbridge.

Ritchie Robert Peel (1899), *The Early Days of the Royal College of Phisitians Edinburgh*, Edinburgh: Johnston.

Rivers, I. (1979), *Classical and Christian Ideas in English Renaissance Poetry*, London: George Allen & Unwin.

Roberts, K.B and Tomlinson, J.D. (1992), *The Fabric of the Body. European Traditions of Anatomical Illustrations*, Oxford: Clarendon Press.

Rogers, J. (1996), *The Matter of Revolution: Science, Poetry and Politics in the Age of Milton*, Ithaca and London: Cornell University Press.

Rosen, E. (trans.) (1967), *Kepler's Somnium: The Dream or Posthumous Work on Lunar Astronomy*, Madison, WI and London: University of Wisconsin Press.

Rueff, J. (1637), *The Expert Midwife, or an Excellent and most Necessary Treatise of the Generation and Birth of Man*, London: Thomas Alchorn.

Sabine, G.H. (ed.) (1965), *The Works of Gerrard Winstanley*, New York, Russell & Russell.

Sarasohn, L. (1984), 'A science turned upside down: feminism and the natural philosophy of Margaret Cavendish', *Huntingdon Library Quarterly*, **47**, 289–307.

Sawday, J. (1995), *The Body Emblazoned: Dissection and the Human Body in Renaissance Culture*, London and New York: Routledge.

Schultz, B. (1985), *Art and Anatomy in Renaissance Italy*, Ann Arbor, MI: UMI Research Press; Epping, UK: Bowker Publishing Company.

Seed, P. (1995), *Ceremonies of Possession: Europe's Conquest of the New World 1492–1640*, Cambridge: Cambridge University Press.

Sennert, D. (1664), *Practical Physick*, London: Peter Cole.

Sermon, W. (1671), *The Ladies Companion, Or The English Midwife*, London: Edward Thomas.

Shakespeare, W. (1986), *Complete Works*, ed. S. Wells and G. Taylor, Oxford: Clarendon Press.

Shapin, S. (1996), *The Scientific Revolution*, Chicago: University of Chicago Press.

—— and Schaffer, S. (1985), *Leviathan and the Air Pump: Hobbes, Boyle, and the Experimental Life*, Princeton, NJ: Princeton University Press.

Sharp, J. (1999), *The Midwives Book, Or the Whole Art of Midwifery*, ed. E. Hobby, New York and London: Oxford University Press.

Shirlaw, L. (1975), 'Dr Archibald Pitcairne and Sir Isaac Newton's "Black Years" (1692–94)' in *The Chronicle of the Royal College of Physicians of Edinburgh*, pp. 23–6.

Shirley, J.W. (1974), 'Sir Walter Raleigh and Thomas Harriot', in Shirley, J.W. (ed.), *Thomas Harriot: Renaissance Scientist*, Oxford: Clarendon Press, pp. 16–35.

Shulman, G.M. (1989), *Radicalism and Reverence: The Political Thought of Gerrard Winstanley*, London, Berkeley and Los Angeles: University of California Press.

Sidney, P. (1962), 'Astrophil and Stella', in *Poems of Sir Philip Sidney*, ed. W.A. Ringler, Oxford: Clarendon Press.

Sigerist, H. (1936), 'The Historical Aspect of Art and Medicine', *Bulletin of the Institute of the History of Medicine*, **4**, 271–97.

Simons, P. (1994), 'Lesbian (In)Visibility in Italian Renaissance Culture: Diana and other cases of *donna con donna*', *Journal of Homosexuality*, **27**, 81–122.

Simpson, E.M. (1924), *A Study of the Prose Works of John Donne*, Oxford: Clarendon Press.

Siraisi, N.G. (1990), *Medieval and Early Renaissance Medicine: An Introduction to Knowledge and Practice*, Chicago: University of Chicago Press.

Slaughter, M. (1982), *Universal Languages and Scientific Taxonomies in the Seventeenth Century*, Cambridge: Cambridge University Press.

Sprat, T. (1667), *The History of the Royal Society of London, For the Improving of Natural Knowledge*, London: n.p.

Srigley, Michael (1988), 'The lascivious metaphor: The evolution of the plain style in the seventeenth century', *Studia Neophilologica*, **60**, 179–92.

Steadman, J. (1971), 'Beyond Hercules: Francis Bacon and the Scientist as Hero', *Studies in the Literary Imagination*, **4**, 3–47.

Sterne, L. ([1760] 1983), *The Life and Opinions of Tristram Shandy, Gentleman*, ed. I.C. Ross, Oxford: Clarendon Press.

Strachey, W. ([1612] 1953), *The Historie of Travell into Virginia Britania*, ed. L.B. Wright and V. Freund, London: Hakluyt Society.

Strong, E.W. (1966), *Procedures and Metaphysics: A Study in the Philosophy of Mathematical-Metaphysical Science in the Sixteenth and Seventeenth Centuries*, Hildesheim: Georg Olms.

Svendsen, K. (1956), *Milton and Science*, Cambridge, MA: Harvard University Press.

Taussig, M. (1993), *Mimesis and Alterity: A Particular History of the Senses*, London and New York: Routledge.

Thomas, K. (1973), *Religion and the Decline of Magic*, Harmondsworth: Penguin.

—— (1983), *Man and the Natural World: A History of the Modern Sensibility*, New York: Pantheon Books.

Thorndike, L. (ed.) (1923), *History of Magic and Experimental Science*, New York: Columbia University Press.

Tobin, T. (ed.) (1972), *The Assembly by Archibald Pitcairne*, Lafayette, IN: Purdue University Studies.

Towler, J. and Bramall, J. (1986), *Midwives in History and Society*, London and Dover, NH: Croom Helm.

Traherne, T. (1932), *Poetical Works*, ed. G.L. Wade, London: Dobell.

Traub, V. (1995), 'The Psychomorphology of the Clitoris', *GLQ*, **2**, 81–113.

—— (1996), 'The Perversion of "Lesbian" Desire', *History Workshop Journal*, **41**, 23–49.

Trye, M. (1675), *Medicatrix, or the Female Physician*, London: Henry Broome and John Leete.

Tuana, N. (1993), *The Less Noble Sex: Scientific, Religious, and Philosophical Conceptions of Woman's Nature*, Bloomington, IN: Indiana University Press.

Vicary, T. (1577), *A profitable Treatise of the Anatomie of mans body*, London: Henry Bamforde.

Vitruvius, M.P. (1960), *The Ten Books on Architecture*, trans. M.H. Morgan, New York: Dover.

Webster, C. (ed.) (1781), *An Account of the Life and Writings of the Celebrated Dr Pitcairne*, Edinburgh and London: Gordon and Murray, Richardson and Urquhart.

—— (ed.) (1970), *Samuel Hartlib and the Advancement of Learning*, Cambridge: Cambridge University Press.

Webster, C. (ed.) (1974), *The Intellectual Revolution of the Seventeenth Century*, London: Routledge & Kegan Paul.

—— (1975), *The Great Instauration: Science, Medicine and Reform 1626–1660*, London: Duckworth.

Wedberg, A. (1955), *Plato's Philosophy of Mathematics*, Stockholm: Almquist and Wiksell.

Wells, N.J. (1989), 'Objective Reality of Ideas in Descartes, Caterus and Suarez', *Journal of the History of Ideas*, **27**, 33–61.

Wilkins, J. (1638), *A Discovery of a World in the Moon*, London: n.p.

—— (1648), *Mathematicall Magick*, London: n.p.

Wodrow, R. (1843), *Analecta or Materials for a History of Remarkable Providences mostly relating to Scotch Ministers and Christians*, Edinburgh: Maitland Club.

Wolley, H. (1674), *A Supplement to the Queen-like Closet*, London: Richard Lownds.

Yates, F. (1964), *Giordano Bruno and the Hermetic Tradition*, London: Routledge & Kegan Paul.

Zamora, M. (1990–91), 'Abreast of Columbus: Gender and Discovery', *Cultural Critique*, **17**, 127–49.

Index

Illustrations are referred to in *italic* type.

For Product Safety Concerns and Information please contact our EU
representative GPSR@taylorandfrancis.com
Taylor & Francis Verlag GmbH, Kaufingerstraße 24, 80331 München, Germany

www.ingramcontent.com/pod-product-compliance
Ingram Content Group UK Ltd.
Pitfield, Milton Keynes, MK11 3LW, UK
UKHW022117190926
475702UK00013B/1489
9781138256439